DATA SCIENCE FOR COMPLEX SYSTEMS

Many real-life systems are dynamic, evolving, and intertwined. Examples of such systems displaying "complexity" can be found in a wide variety of contexts ranging from economics, to biology, to the environmental and physical sciences. The study of complex systems involves the analysis and interpretation of vast quantities of data, which necessitates the application of many classical and modern tools and techniques from statistics, network science, machine learning, and agent-based modeling. Drawing from the latest research, this self-contained and pedagogical text describes some of the most important and widely used methods, emphasizing both empirical and theoretical approaches. More broadly, this book provides an accessible guide to a data-driven toolkit for scientists, engineers, and social scientists who require effective analysis of large quantities of data, whether that be related to social networks, financial markets, economies, or other types of complex systems.

ANINDYA S. CHAKRABARTI is Associate Professor of Economics and UTI Chair of Macroeconomics at the Indian Institute of Management Ahmedabad, India. His main research interests are macroeconomics, big data in economics, time series econometrics, network theory, and complex systems.

K. SHUVO BAKAR is Senior Lecturer at the University of Sydney, Australia. His research interests are Bayesian modeling and computation to reduce uncertainty in inferential statements. He works on statistical machine learning methods and applications to real-life data-driven problems.

ANIRBAN CHAKRABORTI is Dean of Research and Dean of School of Engineering and Technology at BML Munjal University, India. His main research interests lie in the areas of econophysics, data science, quantum physics, and nanomaterial science.

DATA SCIENCE FOR COMPLEX SYSTEMS

ANINDYA S. CHAKRABARTI

Indian Institute of Management Ahmedabad

K. SHUVO BAKAR

University of Sydney

ANIRBAN CHAKRABORTI

BML Munjal University

CAMBRIDGE
UNIVERSITY PRESS

Shaftesbury Road, Cambridge CB2 8EA, United Kingdom

One Liberty Plaza, 20th Floor, New York, NY 10006, USA

477 Williamstown Road, Port Melbourne, VIC 3207, Australia

314–321, 3rd Floor, Plot 3, Splendor Forum, Jasola District Centre, New Delhi – 110025, India

103 Penang Road, #05–06/07, Visioncrest Commercial, Singapore 238467

Cambridge University Press is part of Cambridge University Press & Assessment,
a department of the University of Cambridge.

We share the University's mission to contribute to society through the pursuit of
education, learning and research at the highest international levels of excellence.

www.cambridge.org
Information on this title: www.cambridge.org/9781108844796

DOI: 10.1017/9781108953597

First published 2023

A catalogue record for this publication is available from the British Library.

ISBN 978-1-108-84479-6 Hardback

Cambridge University Press & Assessment has no responsibility for the persistence
or accuracy of URLs for external or third-party internet websites referred to in this
publication and does not guarantee that any content on such websites is, or will
remain, accurate or appropriate.

Contents

Part III Patterns and Interlinkages

Part IV Emergence: From Micro to Macro

Preface

This book grew out of our interactions with the research community working on complex socio-economic systems over the last decade. The literature on complex systems has grown exponentially, trying to answer new questions with a wide array of novel methods. This has ushered in the latest age of research in complex systems where compared to the previous literature, current approaches explicitly rely on data. The focus goes beyond the theoretical confinement of small-scale toy models to tackle empirically documented economic, social, and natural phenomena. As computational cost has drastically reduced over the last couple of decades, it has become easier than ever to conduct statistical and econometric analysis on datasets, small and large alike. This created a chasm as many of the data science toolkits require a deep understanding of the data and, notably, all toolkits are not applicable to all types of data. When we wanted to offer courses on the intersection of data science and complex systems, we faced a barrier. There are many excellent textbooks on different topics pertaining to complex systems as well as data science. But we could not find a single cohesive resource that could provide an overview of the systems and underlying techniques to suit our purpose. This motivated us to bring together many different subject domains under one common theme. The result is this book, which summarizes the relevant streams of the literature and is explicitly geared toward pedagogical usage.

We adopt a data-analysis-oriented approach for an interdisciplinary audience who would be primarily interested in complex systems. In doing so, we not only emphasize computational analysis but also review the latest research problems which are still in scientific infancy. In each chapter, we provide a substantial amount of background material that readers can consult in order to start exploring subject domains independently. Our target audience is advanced undergraduate and graduate students specializing in economics, physics, biology, finance, quantitative sociology, and computer science among others, who are interested in interdisciplinary excursions. Specifically, while writing the chapters, we assumed that the reader would have a good grasp of linear algebra, calculus, and the basics of optimization. Knowledge of basic statistics and probability will be very useful.

We have focused predominantly on complex economic and social systems and described a set of toolkits to analyze them. While the choice of topics is shaped by our own research interests, we have tried to maintain a balance across different domains by providing background and coverage across the topics. We start the technical part of the book by introducing classical and Bayesian statistics along with time series modeling. This part emphasizes inference with non-repeatable experiments based only on observational data, which is more common in economics and the social sciences than in physical sciences. Next, we go through the standard repertoire of machine learning and network theory which is widely used now across multiple disciplines, ranging from social sciences to physical and biological sciences. Perhaps somewhat surprisingly, many of the early-day developments in network theory took place in sociology and economics, although they then found more usage in physics and biology. Over the last decade or so, these toolkits have been very successfully used in most of these subjects. We end the book with a chapter on agent-based modeling which covers six different types of complex behavior related to economic and social systems, with emphasis on applications of statistical as well as computational toolkits.

It would be useful to point out some stylistic features of the work presented here. We start by noting an omission. Nonlinear dynamics was a staple in complex systems literature, which we have deliberately avoided discussing. The reason is that, in our view, nonlinear dynamics has not provided a suitable paradigm to understand economic and social phenomena although there have been many attempts to synthesize such ideas. While things can change in future, at present we consider it to be out of the scope of the book. The second point is about our usage of notations. Since all the topics we discuss here use notations that are fairly standard in the respective subject domains, we have tried to follow the usual notations that are specific to that subject. For example, in statistics it is customary to use μ and σ to describe mean and standard deviation of normal distribution. We do the same unless otherwise explicitly mentioned. But sometimes there are cases where the same notation is used for different objects across disciplines. To account for such instances, we had to deviate from standard practice in some cases and change notations to ensure parity across the chapters within this book. One example that can be highlighted is that i is a convenient notation for lags in time series models, but it is also a standard notation for complex numbers. We have utilized it in both ways and provided clear explanations as to whether we are using it as an index or to denote a complex number.

As we have mentioned above, this book provides a summary of the topics rather than a detailed analysis. For each of the topics, we have mentioned resources which provide a more extensive treatment as well as citing original research papers. All of the results from simulation as well as estimation that have been presented here have been generated using the programming languages Matlab and R. Finally, we note that the book that comes closest to our goal is Thurner et al. (2018), which emphasizes dynamics and information theoretic approach from a statistical mechanics point of view. However, our approach is complementary to theirs in terms of emphasizing socio-economic complex systems and bringing econometrics, statistical inference, and machine learning toolkits to the forefront.

We hope the book will be useful for both researchers and students working on complex systems.

Acknowledgments

We are indebted to our colleagues and collaborators in economics, statistics, and physics – Bikas K. Chakrabarti, Sitabhra Sinha, Arnab Chatterjee, Tathagata Bandyopadhyay, Chirantan Chatterjee, Ratul Lahkar, Shekhar Tomar, Shubhrangshu Manna, Parongama Sen, Bikramaditya Dutta, Tushar Nandi, Thomas Lux, Mauro Gallegati, Tiziana Di Matteo, Taisei Kaizoji, Matteo Marsili, and Sheri Markose. Although they may not necessarily agree with our treatment of the topics or points of view, our many discussions and scientific collaborations with them have shaped the views and thinking that have been expressed in the book. Among many other things, BKC made us aware of the work by astrophysicists Meghnad Saha and Biswambhar Srivastava where they indicated that the Maxwell–Boltzmann distribution from the kinetic theory of gas can be used as a model for the distribution of economic quantities such as income. We found this story fascinating and discuss this in Chapter 6 while describing models of income inequality. We are grateful to our colleagues and collaborators – Soumyajyoti Biswas, Dhiman Bhadra, Indranil Bose, Arnab Chakrabarti, Apratim Guha, Samrat Gupta, Adrija Majumdar, Ankur Sinha, and Vineet Virmani – for carefully reading various chapters of the book and giving us detailed feedback. Without their help, the book would suffer from a lack of nuances in interpretation and many important references. We thank Ankur Sinha for kindly providing the estimated sentiment data from the news items that we have described in Chapter 4. We thank Kiran Sharma and Hirdesh Pharasi for helping us with a few of the schematics in Chapters 4, 5, and 6, and for useful discussions on the initial draft. ASC thanks Jalshayin Bhachech for providing excellent research assistance. We thank our students for giving us feedback on the chapters that were taught as part of coursework. In particular, we have benefited from feedback provided by the students in the postgraduate program on advanced business analytics and the doctoral program at the Indian Institute of Management Ahmedabad. Some of the figures are based on publicly available data. For the rest, we thank the Vikram Sarabhai Library at the Indian Institute of Management Ahmedabad for providing us with the data.

Part I
Introduction

1

Facets of Complex Systems

What is complex about a complex system? This seemingly simple question has neither a simple answer nor a unique one (Ladyman and Wiesner, 2020; Gell-Mann, 2002). Over the past few decades, the science of complex systems has been applied across many disciplines ranging from the physical and biological sciences to the social sciences. Given the widespread usage of the term, an all-encompassing definition is hard to come by. One such definition comes from Rosser (1999) who in turn borrowed the idea from Day (1994): "a dynamical system is complex if it endogenously does not tend asymptotically to a fixed point, a limit cycle, or an explosion." This description squarely relies on the dynamical behavior of a system to categorize it as a complex system. For our purpose, we will take a complementary approach which is considerably broader in scope and phenomenological in its essence (Ladyman et al., 2013). The idea is to not define the system *a priori*, but to describe the characteristics or different facets of complex systems, which may help us to identify a system as complex and to decipher its properties. Loosely speaking, a complex system is one that comprises a *large* number of interacting simple entities, which leads to emergent behavior at the macro level. This emergent behavior of the system cannot be reduced to the behavior of individual entities at the micro level. Complex systems, however different they may be at the micro level, have three broad classes of characteristics associated with their behavior: *heterogeneity*, *interactions*, and *emergence*.

Let us start with a simple example to elucidate the idea of a complex system. Viewed from the angle of literature on complex systems, a wheel is a "simple" entity. A car is a "complicated" object, which represents a collection of many simple entities. And traffic jam is a "complex" phenomenon – literally, not figuratively. The explanation is as follows. A wheel's functionality can be understood by analyzing simple linear dynamics. The functioning of a car, which consists of many parts each of which is simple, can be understood by taking it apart and studying the properties of each constituent part separately. However, the dynamics of a traffic jam cannot be reduced to the motion of each individual car – it is a *complex* system. Instead of looking into how cars behave in isolation, we have to look at how cars interact with each other resulting in a new emergent property called "traffic jam." Such emergent properties are ubiquitous in nature, and they go beyond the realm of human interactions. For example, scientists noted a long time back that how an ant colony functions cannot be understood by observing a single *representative* ant (Kirman, 1993)!

However, the mere existence of a certain type of system may not necessarily imply that it is useful to build a paradigm to study its properties. We also need to know how generalizable the ideas are. It turns out that many phenomena taking place around us do seem to possess the facets of complex systems, at both a theoretical and a practical level. Therefore, a natural question arises: how do we see real-world systems through the lens of complex systems? Real-world systems are dynamic, evolving, and intertwined. Examples of such systems can be found in a wide variety of contexts ranging from the economy to living organisms to the environment to financial markets. The ways in which public opinion swings from one extreme to another, flocks of birds fly in sync, financial markets crash, and languages compete with each other like species with the emergence of dominant languages are all examples of emergent behavior at the scale of underlying systems as a result of interactions among constituent parts.

The next question, therefore, is: how do we make sense of these widely different systems and their behavior? Deciphering complexity is a daunting task as the world around us steadily assumes a more complex and interrelated form. One approach could be from a purely theoretical perspective: for example, by using toy models to establish the so-called "universal" behavior exhibited by some intricate real-world systems. In this respect, sand-pile models exhibiting self-organized criticality have been widely studied. Interacting particle models exhibiting scaling behavior in size distributions are yet another class of models that has become very popular. However, the wide variety of quantitative and qualitative behavior that is demonstrated by complex systems make it seemingly impossible to unify all possible features under one grand theory. Therefore, we follow a different approach. Instead of defining precise universal behavior, which are rare to begin with, we aim to quantify the behavioral traits that describe and/or indicate complexity and build a toolkit to achieve that objective. Fundamentally, the necessary framework to conceptualize a complex system encompasses analytical, computational, and empirical apparatus for studying dynamics that display a wide spectrum of behavior ranging from critical phenomena and complex interactions to the emergence of patterns. Therefore, the challenge appears in the form of approaching the problem with an integrative perspective, in contrast to a reductionist approach.

We pursue an empirical approach toward complex socio-economic systems. The ongoing digital transformation has brought forth new challenges to our current understanding and a new era in studying social and economic systems. A huge amount of data are being generated every moment across different spheres that can be used to understand the complex relationships between the underlying variables. And though the social and economic systems that we study may appear very different from one another, often they have similar characteristics in their dynamics. A crucial aspect of our empirical approach is the explicit attention to phenomenology in the form of a top-down zooming-in approach to deciphering complexity, rather than the prevalent bottom-up approach of describing assumptions and theoretically working out the resulting aggregate implications. In the context of complexity-driven economics, especially macroeconomics, a similar point was made by Di Guilmi et al. (2017), although they differ from us in terms of their emphasis on theory. In the empirical portion of this book, we describe the top-down approach to characterizing macroscopic behavior. Once having built insights into the characterization of macroscopic behavior, we provide a bottom-up approach to arrive at such behavior from microscopic interaction rules similarly to the reductionist approach that connects the behavior of a system with its constituent parts. However, these two approaches retain their differences which arise in the form of emergence. As opposed to the sum of all the smallest component-level behavior representing the macroscopic behavior, we emphasize systems where the macroscopic behavior is more than the sum of its parts.

The nature and features of complexity differ across socio-economic, physical, and biological systems. For quite some time, scientists have gathered sizable amounts of data on physical and biological systems. In recent times, the economy and human society have gone through a revolution due to the explosive growth of information science. This, in turn, is facilitated by rapid technological advances, yielding a high volume and a tremendous variety of data in multiple domains at any given point of time. The big challenge of the present day is in extracting meaningful insights from the data in ways that go beyond data-crunching and data-engineering. More clearly pinned-down rules of interactions governing the dynamics of social and economic agents will lead to a clearer understanding of the corresponding emergent properties. This conjunction is our main thesis.

To explore the emergence of macroscopic behavior from microscopic interactions, we focus on some of the most pressing problems of the present-day world, ranging from social segregation and economic inequality to the extinction of languages, among others. We argue that the paradigm of complex systems can shed light on these phenomena. Clearly, a lot more data are available now to study such systems than was available earlier. We argue that the features of heterogeneity, interactions, and emergence are embedded in each of the systems we examine here, which make the systems complex. With this understanding, we can make sense of the connections between microscopic and macroscopic properties. While this description of complexity is less stringent than other definitions (e.g., see the discussion in Ladyman et al. [2013]), it allows us to simplify our exposition considerably. With that in our view, we will first explain the defining features of complex systems in terms of measurable empirical properties.

1.1 Features of Complex Systems

Some of the early attempts to understand complex systems came from the physics literature and there is a specific reason for that. A natural feature of a complex system is its dynamics, and the study of the mathematics of dynamical systems in the natural world found a home in the physics literature. Such dynamics can be either stochastic or deterministic (including chaotic), often displaying nonlinear and emergent phenomena. To capture system-level dynamics by aggregating the dynamics of all of a system's constituent entities is a non-trivial and often intractable problem. Moreover, the constituents of complex systems are heterogeneous and there can be time-varying interactions between them. Most of the formalisms of statistical physics that are used to understand macroscopic or collective behavior from the dynamics of the microscopic constituents have insufficiencies in this regard. One fascinating example where the application of such theories has been very successful is the type of systems that exhibit self-organization and critical behavior. Notably, the famous "sand-pile" models in physics have been very useful for understanding the universal behavior arising out of such systems. However, as we will discuss later, a direct mapping of such systems on to real-world phenomena has had very limited success. To summarize, the standard approach used in statistical physics has its limitations, although there has been enormous progress in terms of the theoretical understanding of the dynamical properties of complex systems. In order to develop these ideas further, we first need to describe the features of complex systems beyond their dynamics.

The first feature is the stability of an interacting dynamical system. As we have noted above, complex systems can be large and the constituent entities may display heterogeneity. This, in turn, influences the stability of the system. It is probably easiest to grasp this idea with an intuition from ecology. A dominant paradigm in the ecology literature has been centered around the idea that diverse systems would be more stable than their less diverse counterparts. Intuitively, a more heterogeneous system will be able to adapt more to adverse shocks and therefore should display more stability. However, Robert May's work famously turned this idea upside down when he showed that a larger and more diverse system can be actually prone to more instability, rather than stability (May, 1972). The stability of a system also relates to the linkages between its constituent entities. Are more linkages good or bad for stability of a system? May's work showed that more linkages can be bad for stability. A similar insight also carries through in the context of financial markets. More connected financial markets provide more opportunities to diversify and thereby, potentially, create more stability. However, the recent literature recognizes that too many interlinkages may lead to hidden feedback loops, which can destabilize the system rather than imparting stability to it (May et al., 2008; Haldane and May, 2011).

What is the role of interlinkages beyond influencing the stability of a system? To understand this, we need the help of the network view of the system, a useful paradigm for capturing the topological characteristics of a system with non-trivial interlinkages (Newman, 2010). The main understanding comes from the idea that for many (if not most) socio-economic systems, the connections between the constituent entities do not display symmetry and regularity. Connections are heterogeneous and so are their corresponding influences on the system-level properties (Page, 2010). For example, consider a growing online social network. It is dynamic and heterogeneous in its connections across all the users and the diffusion of a piece of news on the network may heavily depend on its topology. More importantly, different social media may actually exhibit non-trivial differences in topology, resulting in differences in macroscopic behaviors such as the diffusion of a piece of news or a rumor. Thus the nature of interlinkages is a crucial component of the complexity of the system.

Competition and the emergence of dominant traits represent another key feature of complex systems (Chakrabarti and Sinha, 2016). One can conceptualize this mechanism via heterogeneity in the macro-behavior of the system. It has long been recognized that even when there is little heterogeneity in core characteristics among the constituent entities of a given system, the outcomes for each of them may diverge widely. An oft-cited example of such a system is income inequality. All characteristics of human beings are possibly well approximated by bell curves – normal distributions, which have a small variance. However, the corresponding income or wealth distribution among the people interacting through the economy has a variance many times larger than the underlying variances in characteristics of the people. The famous Pareto law was first found in income distribution itself, which says that income m is distributed as a power law $p(m) \sim m^{-(1+\gamma)}$ where γ is the Pareto coefficient (typically with a magnitude close to 1, although there are notable exceptions). We will explore this law in much more detail later on when we study size distributions in income and cities. For the time being, one can see the implication of the law in the following way: going by this law with the coefficient being equal to 1, 20% of the entities are responsible for 80% of all events. In this case, that translates into the idea that 20% of the people acquire 80% of the total income. In the context of the present-day world, this degree of heterogeneity and inequality is fairly accurate in an empirical sense although the exact magnitude may differ. Such heterogeneity is seen not only in economic contexts but also in scientific paradigms where citations are power-law distributed and in many other systems that we will describe later. Chakrabarti and Sinha (2016) observed a unique case with movie income distributions, which actually possess bimodal characteristics where success and failure literally translate into two different modes of the outcome distribution. In the extreme, an asymptotic case of inequality in outcomes is seen in the case of language dynamics: some languages may actually die completely, which leads to the emergence of extreme dominance by the remaining languages.

1.2 A Data-Driven View of Complexity

In what follows, we adopt a data-driven view of complex systems. Our focus will be less on the mathematical modeling of complex systems, and much more on the empirical analysis of the resulting behavior of such systems. Recent advances in computation for solving empirical problems have pushed the frontier of science by leaps and bounds. Google's AlphaFold is possibly one of the most impressive achievements in discovering the structures of proteins. This builds on a series of inventions in computer hardware and software over the last century – from the breaking of the Enigma code during the Second World War to IBM's Deep Blue defeating the famous chess grandmaster Garry Kasparov, from IBM's Watson to Google's AlphaZero to the latest technology ChatGPT from OpenAI which can mimic human conversation. This has been possible due to massive advances not only in computational ability but also in data management. At the outset, this may not look like a challenging problem. But often data management turns out to be the biggest practical challenge in computation. One of the great resources has come in the form of shared and distributed computing and data annotation. As we will see in Chapter 4, many learning algorithms require labeled data. Many software packages and websites, among them GitHub open codes, Google Colab, and Amazon Mechanical Turk, have helped the transition from single-user to multi-user crowdsourced computation and data annotation. In what follows, we will not data management any further and will focus almost exclusively on data analysis and modeling.

We consider data collected at different frequencies: low-frequency data (e.g. societal changes spanning centuries), mid-frequency data (e.g. economic data collected at business cycle frequencies in the order of a few years), and high-frequency data (e.g. financial data collected daily or even more frequency). We will refer to such data as time series data. The other dimension of the data would be its heterogeneity. A set of observations at any given point of time would allow us to conceptualize and quantify heterogeneity. We will refer to such data as cross-sectional data.

In order to develop the empirical apparatus, we first provide an introduction to probability and statistics, covering both classical and Bayesian statistics (Chapter 2). We go through a discussion of the classical approach to probability and develop the concept of statistical estimation and hypothesis testing, leading to a discussion of Bayesian models. Then we discuss time series models to analyze evolving systems (Chapter 3). In particular, we develop ideas to model stationary and non-stationary systems. Additionally, we review some ideas from financial econometrics that have proved to be very useful for modeling time-varying conditional second moment: that is, volatility. In the next part of the book, we review machine learning techniques emphasizing numerical, spectral, and statistical approaches to machine learning (Chapter 4). Then we discuss network theory as a useful way to think about interconnected systems (Chapter 5). These four components constitute the building blocks of the data science approaches to complex systems.

1.3 A World of Simulations

There is no single unique framework for studying emergent behavior, and as mentioned above, we often have to borrow tools from several domains such as economics, mathematics, statistics, computer science, and physics. In some sense, the abstract and non-unique nature of this mixed bag of tools is quite useful as one can expand or compress the scope of analysis as necessary.

Due to the property of emergence, for many of the complex systems and their dynamical patterns, characterization of local rules may not directly capture systemic behavior. Whether we are studying self-organization in economic markets, the dynamics of ethnic conflicts and cooperations, migration networks, or the spread of epidemics, the modeling approach has to emphasize the varying scales and nature of interactions. However, there is another dimension to this problem. Along with making predictions for the expected behavior of the system, one may also want to see what are the possible extreme cases. More generally, one may want to first figure out what are the boundaries of possible behavior of a given system. This is very important for systems that are prone to sudden collapse. Markets, in particular financial markets, provide a great example of such behavior. For policymakers or even for market participants, a better understanding of large fluctuations is typically more useful, for hedging against extreme outcomes, than the prediction of average behavior.

How can we delineate the boundaries of all possible behaviors of a system consisting of a large number of interacting entities, each with potentially non-trivial dynamics? This poses a big challenge. Analytically enumerating all possible behaviors is a daunting task and often impossible for all practical purposes. The complex systems literature has taken an alternative path to solving this problem. Rather than trying to do brute-force calculations by ourselves, it is easier to ask the computer to do so given the cheapness of computational power. The idea is to simulate possible behaviors with simple iterative rules governing interactions between individual agents, which can be tuned to mimic the dynamics of emergent phenomena. Additionally, this approach is useful for finding the average behavior of systems which are analytically intractable.

In order to develop insights into real-world socio-economic systems, in the later part of the book we describe six models that illuminate different facets of complex systems. We start with a model of social segregation which introduces a spatial model of interaction between agents following the seminal work by Schelling (1971). This is an elementary model that delivers the astounding insight that even a minor preference for a homogeneous neighborhood in a heterogeneous population may quickly result in strong segregation of the population. At the same time, this model also started the literature on interaction models of grids or lattices, connecting it to the idea of cellular automata. Famous examples of this kind of model include Conway's Game of Life model and cellular automata model. Interested readers may consult Wolfram (1984, 2002) for an in-depth view of such systems in the context of discrete mathematics. Next, we describe a more intricate scaling behavior on a square lattice in the form of the Bak–Tang–Wiesenfeld model (Bak et al., 1987; Bak, 2013) that exhibits self-organized criticality through the famous simulated sand-piles. A follow-up description of the Bak–Sneppen model sheds light on extinction through competition in a multi-agent setup. In particular, this model generates the scaling behavior of avalanches which models extinctions. A simple time series analysis shows periods of bursts and calm, capturing *punctuated equilibrium* behavior – reminiscent of the theory proposed by Gould and Eldredge (1977). Building on these abstract spatial models, next we focus on city size distribution which possesses a well-known defining feature in the form of Zipf's law, a special case of the Pareto law described above. This law provides a natural restriction on the spatial organic growth of cities. We study two different mechanisms that explain the appearance of Zipf's law. Then we introduce bilateral interaction between agents to explain the dispersion of agent-level attributes. As a modeling paradigm, we then take up models of asset inequality. Empirically, such inequality seems to exhibit robust behavior across economies and time especially in the right-hand tails of income and wealth distributions, in the form of Zipf's law. We show that heterogeneity in general and Zipf's law in particular arise out of interactive systems. Next, we consider competition across groups of agents. As a modeling paradigm, we take up the case of linguistic competition where the emergence of dominant and recessive languages occur as an outcome of competition. Until this point, all of the models we have discussed can be thought of as representing interactions between zero-intelligence agents. In the last model, we consider agents with limited intelligence or rationality who can strategically compete against each other. We analyze a resource competition game to study the evolution of coordination and anti-coordination. We close the discussion with a brief detour through the trade-off between the realism and the generalizability of these models and corresponding insights.

Part II
Heterogeneity and Dependence

2

Quantifying Heterogeneity
Classical and Bayesian Statistics

A defining feature of complex systems is heterogeneity. From ant colonies to human behavior on social media to the distribution of wealth among people, systems differ in the degree of heterogeneity of traits and outcomes. This observation leads to two questions. First, how to quantify this heterogeneity? Second, if we can quantify the heterogeneity, does it reveal any underlying properties of the system? From a modeling perspective, both are pertinent questions. Any system of interacting entities would follow an underlying mechanism of interaction which leads to the macroscopic behavior observable to the modeler. Often the goal of the modeler is to infer this underlying mechanism by looking at the behavior of the system.

For studying real-world systems, a natural first step would be to systematically analyze data sampled from those systems. At an elementary level, simple characterization in terms of average and dispersion from a given sample of observed data may give us a crude idea about the broad features of the system. However, we also want to understand the underlying *data-generating processes*. Knowing about the data-generating process helps us in making inferences about the process as a whole rather than the specific sample we observe. This in turn feeds into understanding the true underlying mechanism that gives rise to the observed behavior.

Statistics is a discipline that concerns studying data with the goal of extracting meaningful information from some observed phenomena. Broadly speaking, the discipline has two approaches: classical or frequentist and Bayesian. The approach of classical statistics provides the most elementary introduction to the ideas in statistics. In contrast, the Bayesian school (Bayes and Price, 1763) of statistical concept was developed to solve inferential problems (Efron, 1986), where existing beliefs about a given system are coupled with observed data to obtain updated beliefs about the system. Statistics literature in the nineteenth century was dominated by the Bayesian approach (Efron, 2005). In the twentieth century, R. A. Fisher ("the single most important figure" in statistics as described by Efron [1998]) bridged the Bayesian and frequentist or classical viewpoints. Fisher and his contemporary statisticians in the first half of the twentieth century pointed out that making inferences in the Bayesian framework would lead to practical problems requiring heavy and complicated computations. Such computations were impossible to carry out in those days in the absence of modern computing power. This led to the Neyman–Pearson–Wald era of decision theory, roughly from the 1930s to the 1950s, when the foundation of classical statistics was developed (Efron, 1986). In this framework, heavy computations were avoided by imposing theoretical assumptions resulting in the flourishing of the frequentist approach. Over the ensuing few decades, there was a race between these two approaches to become the superior paradigm. However, the debate surrounding their relative superiority has softened in recent years with the understanding that both paradigms can offer complementary solutions to statistical problems. Often they have relative advantages without complete domination of one approach over the other (Bayarri and Berger, 2004).

In this chapter, we discuss the fundamentals of statistics and usage of the apparatus for solving problems through both classical and Bayesian paradigms. To relate to the characterization of heterogeneity in complex systems, this chapter begins with a discussion on probability distributions in Section 2.2 following a brief overview on data processing in Section 2.1, where we discuss and explain some common terms and definitions. In Section 2.3, we discuss the classical way of thinking about statistics with some definitions and related theorems. Section 2.4 describes the Bayesian approach and gives some popular computational methods for solving problems. Finally, Section 2.5 discusses some multivariate statistical methods. Throughout this chapter, we have followed standard definitions and theorems. Unless specified otherwise, we follow standard notation used in the literature. We provide a few examples along with the theories in order to work out some of the implications of those theories. We have provided an extensive list of references to books and journal articles so that the reader can follow up in more detail.

Before getting into the discussion, it is useful to clarify a specific usage of the term *model complexity* (sometimes we use the term *complexity of a model*) which is used to denote the number of variables in a model and the way they interact with each other. This is not the same as the concept of *complexity* in the sense of complex systems. We stick to the usage of this term in this chapter as the literature has used it frequently.

2.1　Data Characteristics

The foremost feature of data generated by complex systems lies in the *dependence* and *interdependence* of variables. We define the term *variable* as a characteristic of data that can be measured. Such characteristics can be from anywhere. For example, a dataset with information on surface air temperature represents environmental

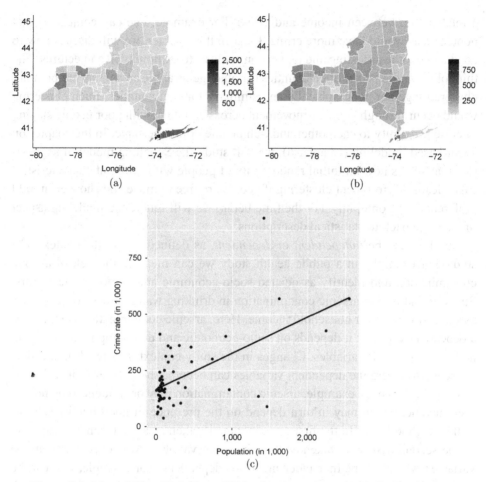

Figure 2.1 Map of New York State, USA, showing county-level distributions of population and crime (the bars in grayscale represent the variable in 1,000s) in panels (a) and (b), respectively; a scatterplot of the two variables with a linear fit is shown in panel (c) (data obtained from the Division of Criminal Justice Services, New York).

conditions. A dataset with information on households income represents economic conditions. A dataset with information on stock prices represents financial conditions of the corresponding company.

As an example of socio-economic data, consider the map of the US state of New York showing county-level distribution of population and crime (Figure 2.1). The data came from the Division of Criminal Justice Service, New York. We make three observations here. The first observation is general in its scope. Inclusion of more information leads to potentially better understanding of any given system. However, this comes at the cost of extracting meaningful insights from larger and more diverse datasets. Second, in this particular case, one can wonder whether there is a relationship between income and crime. For example, one can conjecture that poorer areas would have more crime. Later in this chapter, we will discuss how to conduct hypothesis testing and regression analysis to explore such conjectures. The third observation is that often counties (or other geographic areas) that are poor are clustered together geographically and similarly for rich counties. Such clustering would occur through people's movement across counties, with poor people staying in close proximity to each other and rich people doing the same. In the chapter on agent-based modeling (Chapter 6), we will study the Schelling model of segregation that shows how an initial random mix of people with different characteristics easily leads to prominent clustering if people are free to move and show even mild preference for homophily. For the time being, we will ignore the modeling aspect of it and go back to statistical descriptions.

Variables can be *independent* or *dependent*, as defined within the context of a study. For example, in a public health study we can measure the risk of arsenic contamination and identify associated socio-economic and demographic factors. Suppose that higher arsenic contamination in drinking water across households is associated with lower household income. Here, arsenic contamination is treated as a dependent variable: it depends on socio-economic and demographic factors, that is, the independent variables. Changes in the study context may result in a different scenario where the dependent variables can become the independent variables. Expanding the above example, arsenic contamination may be influenced by household income, which may in turn depend on the presence of good-quality schools in the neighborhood. In the first case, income is taken as an independent variable. In the second case, it is taken as the dependent variable. Moreover, there can be variables which are neither independent or dependent. For example, one can be interested in identifying groups of observations where variables are closer to each other in magnitude within groups vis-à-vis across groups.

We can group variables into *metric* and *non-metric* types. In a *metric* system, the variables are *continuous* or *quantitative* in nature. In a *non-metric* system, the variables are mainly treated as *categorical*. For example, air temperature measured on a continuous scale is a *metric* variable, whereas a variable reflecting gender (say male or female or otherwise), capturing distinct classes or categories, can be treated as a *non-metric* variable.

Finally, we define data characteristics based on their collection system. For example, variables in a dataset measured or collected for multiple entities at a single time point is called *cross-sectional*, whereas *time series* data denotes a variable for which the data is collected at multiple time points. A richer type of data comes in the form of *panel* or *longitudinal* data, where each variable is observed over more than one time period, giving multiple entities that are observed at multiple time points.

2.1.1 Data Processing

The first step of data analysis is processing the data and describing it via summary statistics. This step provides guidance for solving the problems at hand as well as defining the boundary of the problem to be analyzed. From a statistical perspective, appropriate data processing can be divided into two broad classes:

1. by describing (e.g. summarizing and interpreting numbers or characters that we are referring to as data) and
2. by identifying patterns (e.g. distributional patterns, regularities/irregularities, network/hierarchic natures) for the purpose of making inferential decisions that take into account the measurement of uncertainty.

Given a dataset, we will describe it as a sample. Suppose n number of observations are given to us. These observations constitute a "sample." We will discuss the origin of this sample later. For the time being, let us imagine that the only available information is about the values of the observations (say, county-level rates of crime and population in New York as shown in Figure 2.1). We can imagine that each observation corresponds to an underlying random variable, leading to the idea that the sample would correspond to a collection of n random variables: $X_1, \ldots, X_n$.

A *population* is a collection of all random variables describing some behavior of a system under study. Theoretically, this includes all information about the specific behavior of the system that the modeler is interested in. In contrast, a *sample* is a collection of random variables drawn from the population. Ideally, it should reflect (in an inferential sense) the properties of the whole population in a smaller but representative subset of the population. Statistical inference allows us to conjecture about a population based on observation of a sample. Measures of given characteristics describing a population are known as *parameters*. If such a measure is obtained from a sample, then we call it a *statistic*.

We will discuss the descriptive measurement of statistics in subsequent sections, and later in this chapter we will discuss inferential statistics in detail.

Measures of Central Tendency and Dispersion

Descriptive statistics commonly explain data using measures of *central tendency* and *dispersion*. Three standard measures of central tendency are: (i) arithmetic mean (in the rest of this chapter we denote arithmetic mean as simply "mean"), (ii) median, and (iii) mode. The most common measures of dispersion are: (i) range and (ii) variance (or standard deviation).

Before explaining central tendency and dispersion, let us define the term *frequency*, which tells us how often a given value occurs for a particular variable.

For example, in a pharmacy we have eight of sales records for a particular anti-inflammatory drug. The variable X, the number of units sold, is recorded as $3, 4, 4, 5, 5, 5, 7, 8$ units. Suppose we ask how often 5 units of the drug are sold. As can be seen, the answer is 3 as the number 5 occurs 3 times within the 8 days. Therefore, the relative presence of this particular value is 3/8. This is the basic idea of frequency of a particular value of a variable X.

Now we can describe some basic measures. The sample mean is described as $\bar{X} = \frac{1}{n}(X_1 + \cdots + X_n) = \frac{1}{n}\sum_{i=1}^{n} X_i$. The median is the second measure of central tendency, which splits the observations into two parts with (roughly) the same number of observations in each of them. Mathematically, we rank-order the observations. When n is odd, we define median as the $\left(\frac{n}{2} + 0.5\right)$th term; when n is even, we define it as the average of the $\frac{n}{2}$th and $\left(\frac{n}{2} + 1\right)$th terms. The mode is the observation that has maximum frequency. We write range as the difference between maximum and minimum observation of the population, that is, $X_{max} - X_{min}$. We define variance as $\frac{1}{n}\sum_{i=1}^{n}(X_i - \bar{X})^2$. A square-root transformation of variance yields the standard deviation. The population counterparts of mean and variance (the *true* values) are often denoted by μ and σ^2. The idea of statistical inference is to estimate the true values of the population's characteristics (like μ and σ) by observing only sample's analogues of the same (like $\bar{X}$ and s). Later on we will discuss the concept of an unbiased estimator. For the time being, we note that $s^2 = \frac{1}{n-1}\sum_{i=1}^{n}(X_i - \bar{X})^2$ with the denominator for sample variance being $n - 1$ (not n) representing the *degrees of freedom* in the sample. This is an unbiased estimator of the population variance (we will describe unbiasedness in Section 2.3).

Example 2.1 Consider the above example, where we have sales records of $3, 4, 4, 5, 5, 5, 7, 8$, and assume that these are population observations. Hence, the mean is 5.125, the median is 5, and the mode is 5. We can also see that the maximum and minimum values are 8 and 3, respectively; thus the range is 5, the variance is 2.696, and the standard deviation is $\sqrt{2.696} = 1.642$.

Five-Number Summary and Coefficient of Variation

The five-number summary represents minimum, maximum, median, first quartile, and third quartile of a set of observations which are rank-ordered. Quartiles divide the whole set of rank-ordered observations into four parts, where the first quartile represents the 25% mark and the third quartile represents the 75% mark. Note that the second quartile, that is, the 50% mark, is the median. Graphically, we can present the five-number summary with the help of a box and whisker plot (see, for

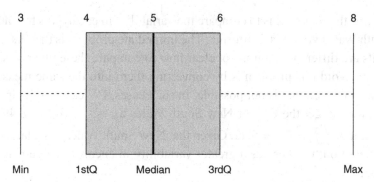

Figure 2.2 Example of box and whisker plot representing the five-number summary.

example, McGill et al. [1978] for a classic reference). Figure 2.2 shows an example that we explain in this section.

Example 2.2 Consider the set of rank-ordered observations $3, 4, 4, 5, 5, 5,$ $7, 8$. We write the five-number summary as: minimum 3, maximum 8, median 5, first and third quartiles 4 and 6, respectively. The corresponding box and whisker plot is given in Figure 2.2.

The coefficient of variation (CV) is a measure often used to compare variability between two sets of observations without conducting any statistical tests. We define the population CV as:

$$CV = \frac{\sigma}{\mu} \times 100, \tag{2.1}$$

where μ is the mean and σ is the standard deviation of a population. Similarly to the population CV, we write the sample CV as $CV = \frac{s}{\bar{X}} \times 100$.

Example 2.3 The CV shows the relative variability around the mean. This is important if the two series of observations have different measurement units. For example, suppose we have a series of sample observations measured from a LiDAR (light detection and ranging) image (Dubayah and Drake, 2000) of eucalyptus trees in a 50-square-meter area in the hills of the Great Dividing Range in New South Wales, Australia. The average height of the trees is 52 meters with a variance of 18. In another sample from California, USA, it reported that in a similar 50-square-meter area the average height of eucalyptus trees is 142 feet with a variance of 55.

Let's assume that our goal is to compare the variability in eucalyptus tree heights in New South Wales vis-à-vis California. The immediate problem is that the measurement units are different and it is not clear how to compare these two datasets. One way we can avoid this problem is by converting them into the same measurement unit. However, often that is not possible. In such cases, CV can be of help. For our example, we can get the CV for New South Wales as $\frac{\sqrt{18}}{50} \times 100 = 8.49$ and for California, it is $\frac{\sqrt{55}}{142} \times 100 = 5.22$. Given that New South Wales has a larger CV, we can say New South Wales has a greater variability in eucalyptus tree heights than California.

2.2 Probability Distributions

2.2.1 Intuitive Idea

Here we will describe the concept of probability in an informal way. Probability captures the idea of "chance." Consider tossing a fair coin. It has two possible outcomes: heads or tails. Since the coin is fair, the chance of getting heads is $\frac{1}{2} = 0.5$, and similarly the chance of getting tails is 0.5. Intuitively, this captures the idea of probability – the chance of the occurrence of events or the outcome of an experiment. In the coin-tossing example, the events are appearances of heads or tails. For the sake of completeness, we note that the example of a fair coin to explain probability or chance is somewhat tautological as the definition of a fair coin itself presupposes the concept of probability. A complete description of probability requires a much larger and deeper range of mathematical tools, which we will omit here (see Ross [2014] for a more detailed exposition; see Billingsley [2008] for a fully measure-theoretic exposition). For our current purpose, we will stick to the same example due to its relative ease of exposition.

Building on the coin-tossing example, consider a more general experiment with multiple outcomes, which is conducted repeatedly. Let us first define the set of all possible outcomes of the experiment. We will call this set the "sample space." Let us denote it by S. A possible outcome of the experiment will be called an "event." Mathematically, we will denote it by subset Δ of S. Let's denote the probability of an event Δ by $P(\Delta)$. This $P(.)$ follows three axioms: (i) $0 \leq P(\Delta) \leq 1$, (ii) $P(S) = 1$, and (iii) if Δ_1 and Δ_2 are disjoint then $P(\Delta_1 \cup \Delta_2) = P(\Delta_1) + P(\Delta_2)$. The last property is called finite additivity. More generally, $P(.)$ satisfies additivity with a countable number of sets.

Remark Axiom (i) is a normalization which requires that probability values must be between zero and one. Axiom (ii) states that the probability is one for all possible outcomes S of the experiment. Axiom (iii) states that if events Δ_1 and Δ_2

are mutually exclusive, that is, both cannot take place at the same time, then the probability that either event Δ_1 takes place or Δ_2 takes place is equal to the sum of their corresponding probabilities. On the other hand, if events Δ_1 and Δ_2 are not mutually exclusive, then we can write the full expression as $P(\Delta_1 \cup \Delta_2) = P(\Delta_1) + P(\Delta_2) - P(\Delta_1 \cap \Delta_2)$.

The probability of an event Δ_2 given that another event Δ_1 has occurred is given by the following conditional probability:

$$P(\Delta_2|\Delta_1) = \frac{P(\Delta_2 \cap \Delta_1)}{P(\Delta_1)}$$

where $P(\Delta_1) > 0$. This is sometimes described as the "forward-looking" probability which evolves from Δ_1 event to Δ_2. If two events Δ_1 and Δ_2 are independent, we can write $P(\Delta_1 \cap \Delta_2) = P(\Delta_1)P(\Delta_2)$. Therefore, we can write the conditional probability in such cases as

$$P(\Delta_2|\Delta_1) = \frac{P(\Delta_2 \cap \Delta_1)}{P(\Delta_1)} = \frac{P(\Delta_2)P(\Delta_1)}{P(\Delta_1)} = P(\Delta_2).$$

Remark Conditional probability leads to the concept of Bayes' theorem which allows us to identify an event through a "backward-looking" approach. Suppose we want the probability evaluated in the opposite direction, that is, to obtain the probability of an event Δ_1 given that we know Δ_2 and we consider $P(\Delta_2)$ as prior information. Then we can write the basic statement of the Bayes' theorem as

$$P(\Delta_1|\Delta_2) = \frac{P(\Delta_1 \cap \Delta_2)}{P(\Delta_2)} = \frac{P(\Delta_1)P(\Delta_2|\Delta_1)}{P(\Delta_2)}. \tag{2.2}$$

We can expand the denominator in the following way.

Let us consider the *law of total probability* –

$$\begin{aligned} P(\Delta_2) &= P(\Delta_1 \cap \Delta_2) + P(\Delta_1' \cap \Delta_2) \\ &= P(\Delta_1)P(\Delta_2|\Delta_1) + P(\Delta_1')P(\Delta_2|\Delta_1'), \end{aligned} \tag{2.3}$$

where Δ_1' is the complement of event $\Delta_1 \subseteq S$. More generally, we can write $P(\Delta_2) = \sum_\Delta P(\Delta \cap \Delta_2) = \sum_\Delta P(\Delta)P(\Delta_2|\Delta)$ where the subsets Δ constitute a partition of the set S. This we can substitute in Equation 2.2 which in turn leads to an elaborate version of the Bayes' theorem:

$$P(\Delta_1|\Delta_2) = \frac{P(\Delta_1)P(\Delta_2|\Delta_1)}{\sum_\Delta P(\Delta)P(\Delta_2|\Delta)}. \tag{2.4}$$

This theorem is tremendously important as it forms the backbone of Bayesian statistics. We will pick it up again in Section 2.4.

Example 2.4   Consider a scientist experimenting with a method for slowing down the spread of a ribonucleic-acid-based (RNA-based) virus causing fever, with the help of contact-tracing in a community. The scientist considers the following key steps for contact-tracing of the patients: identification of the patient, identification of the mode of transmission (e.g. close contact with someone who has already contracted the virus), duration of the infection, and the period of self-isolation or self-quarantine. There is another step where the scientist wants to also consider the individuals who are asymptomatic, who are often a trigger for spreading the virus, and who are hard to identify.

After doing some research and collecting data from 59 patients, the scientist gets some initial results. To begin with, 15 patients do not know how they got infected by the virus. To relate it to the sample space, we have two events – (i) patients know how they became infected and (ii) patients don't know the source of their infections. Thus, $P(A) = (59 - 15)/59 \approx 0.75$ and $P(A') = 15/59 \approx 0.25$, where A and A' represent the events "known source" and "unknown source," respectively. The scientist also finds that 43 patients who knew the source of infection self-quarantined for 14 days. However, six patients failed to self-quarantine among the patient group who did not know the source of infection.

The scientist is interested in identifying the probability that a patient self-quarantined (call it event Q) given that the infection source is known (corresponding to event A). This probability would be

$$P(Q|A) = \frac{43}{(59 - 15)} \approx 0.98.$$

Similarly, the probability that a patient self-quarantined (Q) given that the infection source is unknown (A') is written as

$$P(Q|A') = 1 - P(Q'|A') = 1 - \frac{P(Q' \cap A')}{P(A')} = 1 - \frac{6}{15} = 0.6.$$

Now to get the probability that a random patient self-quarantined (Q) who also has an unknown source of infection (A'), we can find

$$P(A' \cap Q) = P(A') \times P(Q|A')$$
$$= \frac{15}{59} \times \frac{15 - 6}{15} \approx 0.25 \times 0.6 = 0.15.$$

Further, suppose the scientist is interested in finding the probability that a patient doesn't know the source of infection (A') given that they self-quarantined (Q). This reflects the idea of "backward-looking" calculation where the scientist tries to understand patients' information about the unknown infection source given that

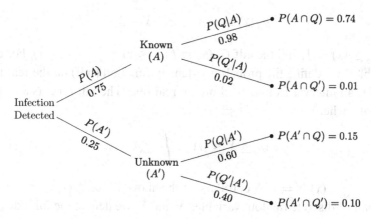

Figure 2.3 Tree diagram showing various unconditional and conditional probabilities arising out of detection of infection.

the patient has self-quarantined. Here, we can apply Bayes' theorem to get the probability:

$$P(A'|Q) = \frac{P(A') \times P(Q|A')}{P(A') \times P(Q|A') + P(A) \times P(Q|A)}$$

$$= \frac{0.25 \times 0.6}{0.25 \times 0.6 + 0.75 \times 0.98} \approx 0.17.$$

Figure 2.3 shows a tree diagram of this example.

Distribution Function

A distribution function, also known as the cumulative distribution function (cdf), is a function generated from the probability distribution, accumulating up to particular values of a random variable. Formally, for a random variable X, a cdf is written as

$$F_X(x) \equiv P(X \leq x) \quad x \in \mathbb{R}. \tag{2.5}$$

If x_1 and x_2 are two arbitrary numbers where $x_1 < x_2$, we have $P(x_1 < X \leq x_2) = F_X(x_2) - F_X(x_1)$.

A random variable X could be discrete or continuous. A discrete variable can take a finite or countably infinite number of values. In contrast, a continuous variable is not countable but we can measure it. Discrete probability distributions are constructed from discrete random variables and can be expressed by a probability mass function (pmf):

$$g_X(x) \equiv P(X = x) \quad x \in \mathbb{R} \tag{2.6}$$

where $\sum_x g_X(x) = 1$, and the cdf $G_X(x) \equiv P(X \le x) = \sum_{x:z \le x} g_X(z)$. For continuous variables, we define the probability density function (pdf) on the real line. Let us consider an interval from α to β on the real line. The pdf $f_X(x)$ is a continuous function of x where

$$P(\alpha \le X \le \beta) = \int_\alpha^\beta f_X(x)dx \tag{2.7}$$

such that $\int_{-\infty}^\infty f_X(x)dx = 1$. A corollary of the above is that $f_X(x) = \frac{d}{dx}F_X(x)$.

For a pair of discrete random variables X and Y, we denote the joint distribution function by $g_{XY}(x, y)$ where $g_{XY}(x, y) = P(X = x, Y = y)$. The cdf for (X, Y) is written as $G_{XY}(x, y) = P(X \le x, Y \le y)$ and the corresponding conditional probability function of X given Y is written as $g_{X|Y}(x|y) = \frac{P(X=x,Y=y)}{P(y=y)}$ where $P(y = y) > 0$. We write the marginal probability mass function of X as $g_X(x) = \sum_y g_{XY}(x, y)$. For continuous variables, analogous definitions will go through with the summation signs being appropriately interchanged with integration signs.

2.2.2 Mathematical Expectation

Let X be a random variable with probability mass function $g_X(x)$ (for discrete variable) or density (for continuous variable) function $f_X(x)$. If the expectation of X exists, we write it as

$$E(X) = \begin{cases} \displaystyle\sum_{x \in S} x g_X(x) & \text{for discrete and} \\ \displaystyle\int_{-\infty}^\infty x f_X(x)dx & \text{for continuous variable } X. \end{cases} \tag{2.8}$$

We write the corresponding variance as

$$Var(X) = E([X - E(X)]^2) = E(X^2) - E(X)^2. \tag{2.9}$$

Remark Some useful properties of expectation include the following. It is additive: $E(X + Y) = E(X) + E(y)$ and $E(aX + b) = aE(X) + b$ where a and b are some constant terms. Variance has slightly different properties. First, $Var(aX + b) = a^2 Var(X)$ as the variance of a constant term is zero ($Var(b) = 0$). In terms of additivity, we can write the expansion as $Var(X + Y) = Var(X) + Var(y) + 2Cov(X, Y)$ where $Cov(X, Y)$ is the covariance of X and Y. We will define covariance in the next paragraph. A very useful inequality involving the expectation operator comes in the form of the Cauchy–Schwarz inequality – $[E(XY)]^2 \le E(X^2)E(y^2)$.

Example 2.5   Consider a gamble with a slightly biased coin with probabilities –
P(heads) $= 0.47$ and P(tails) $= 0.53$. Let the price to play the gamble be INR 10
(INR represents Indian currency – $1 corresponded to around INR 76 on March
16, 2022). Suppose that if heads appear one gets INR 20 and if tails appear one will
lose INR 10. The expected gain from tossing the coin once is $20 \times 0.47 + (-10) \times$
$0.53 = 4.1$. Thus the expected loss from playing this gamble is INR $(10 - 4.1) =$
INR 5.9.

Let X and Y be two random variables. We denote the covariance between X and
Y by

$$Cov(X, Y) = E[(X - E(X))(Y - E(Y))]$$
$$= E(XY) - E(X)E(Y). \tag{2.10}$$

The second equality follows from the properties of expectation. If $Cov(X, Y) \neq 0$,
the variables X and Y are not independent. It is useful to note that this relationship holds only in one direction – independence implies zero covariance, but
zero covariance does not imply independence (higher-order covariances may be
present).

A more common measure of interdependence is the correlation coefficient which
can be obtained from covariance using the following expression:

$$\rho_{XY} = \frac{Cov(X, Y)}{\sqrt{Var(X)}\sqrt{Var(Y)}}. \tag{2.11}$$

Remark   The measures of covariance and correlation provide an idea about the
extent of linear relationship between two variables. A correlation value equal to 1
implies that the corresponding variables are perfectly positively correlated, whereas
-1 refers to a perfectly negative correlation between them. Zero magnitude implies
that the variables are not correlated. Note that the way correlation is calculated
here represents a linear relationship between the variables. It is possible that two
variables have a nonlinear relation but still produce zero correlations. For example,
consider a scatter plot where Y is distributed along a sin curve on the X-axis.

2.2.3 Moments

Moments represent important characteristics of a distribution in terms of parameters underlying the population. We have already noted earlier that $E(X)$ and
$Var(X) = E([X - E(X)]^2)$ represent the mean and variance of a random variable X.

These two measures are known as the first and the second moments, respectively. We can generalize the idea further to higher orders.

Let r be a positive integer and X be the random variable with pmf $g_X(x)$ or pdf $f_X(x)$. We define the rth moment about the origin as

$$\mu'_r = E(X^r) = \begin{cases} \sum_{x \in S} x^r g_X(x) & \text{for discrete variables,} \\ \int_{-\infty}^{\infty} x^r f_X(x)dx & \text{for continuous variables.} \end{cases} \qquad (2.12)$$

Remark The rth moment about mean $\mu_r = E([X - \mu]^r)$ where $\mu = E(X)$. It can be shown with a little bit of algebra that the following equalities hold:

$$\mu_2 = \mu'_2 - \mu'^2_1,$$
$$\mu_3 = \mu'_3 - 3\mu'_2\mu'_1 + 2\mu'^3_1,$$
$$\mu_4 = \mu'_4 - 4\mu'_3\mu'_1 + 6\mu'_2\mu'^2_1 - 3\mu'^4_1,$$

and so on.

Skewness and Kurtosis

We define skewness as a measure that quantifies the lack of symmetry in a distribution. For an asymmetric distribution, we can identify asymmetry either on the right or left tails of the distribution. If the mean of a distribution is greater than mode, then the skewness is positive. If the mean is less than the mode, then the distribution has a negative skewness. We write the moment-based measure of skewness with Pearson's coefficient $\gamma_1 = \frac{\mu_3}{\sigma^3}$ where σ is the standard deviation. Two other ways we can define skewness based on mode or median of the distribution are as follows:

$$\gamma_{\text{mode}} = \frac{E(X) - \text{Mode}(X)}{\sqrt{Var(X)}},$$

$$\gamma_{\text{median}} = \frac{3 \times (E(X) - \text{Median}(X))}{\sqrt{Var(X)}}. \qquad (2.13)$$

A zero value of γ_{mode} or γ_{median} implies that the distribution is symmetric. A negative value of this parameter indicates that the distribution is negatively skewed and a positive value indicates that the distribution is positively skewed.

In a distribution, kurtosis measures the degree of peakedness or its opposite, flatness. The moment-based measure of kurtosis is $\beta_2 = \frac{\mu_4}{\mu_2^2}$ and the corresponding coefficient is $\gamma_2 = \beta_2 - 3$. There are three types of distributions in terms of kurtosis: (i) platykurtic, (ii) leptokurtic, and (iii) mesokurtic. The platykurtic shape (also known as $-$ve kurtosis) is flat and spread out with coefficient $\gamma_2 < 0$, whereas the leptokurtic shape (also known as $+$ve kurtosis) is high and thin with $\gamma_2 > 0$. The

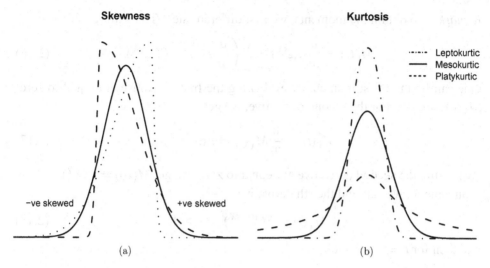

Figure 2.4 Probability density functions with different types of (a) skewness and (b) kurtosis.

mesokurtic shape has a balance with not too much flatness or peakedness. This has the coefficient $\gamma_2 = 0$ (e.g., Gaussian distribution). Figure 2.4 shows density functions related to (a) skewness and (b) kurtosis.

Moment-Generating Function

For some distributions, it may be challenging to find the moments directly. In such cases, moment-generating functions (mgf) can be used. The nature of such a function is that successive differentiation leads to a sequence of moments.

Let us define $M_X(t)$ as the moment-generating function of a random variable X for all $t \in \mathbb{R}$. We write

$$M_X(t) = E(e^{tX}) = \begin{cases} \sum_x e^{tx} g_X(x) & \text{for discrete variables,} \\ \int_{-\infty}^{\infty} e^{tx} f_X(x) dx & \text{for continuous variables.} \end{cases} \tag{2.14}$$

We can write $M_X(t)$ by expanding the exponential series:

$$M_X(t) = E(e^{tX}) = 1 + tE(X) + \frac{t^2}{2} E(X^2) + \cdots$$

$$= 1 + t\mu_1' + \frac{t^2}{2}\mu_2' + \cdots$$

$$= \sum_{r=0}^{\infty} \frac{t^r}{r} \mu_r' \tag{2.15}$$

where μ_r' is the rth moment about origin.

Remark   To obtain the moments, we first differentiate $M_X(t)$ with respect to t:

$$M'_X(t) = \frac{\partial}{\partial t}E(e^{tX}) = E\left(\frac{\partial}{\partial t}e^{tX}\right) = E(Xe^{tX}). \qquad (2.16)$$

One can find the first moment by evaluating the first derivative at t equal to zero: $M'_X(0) = E(X)$. For the second derivative, we get

$$M''_X(t) = \frac{\partial}{\partial t}M'_X(t) = E(X^2 e^{tX}). \qquad (2.17)$$

Evaluating the second derivative at t equal to zero, we get $M''_X(0) = E(X^2)$.

In general, we can find the rth derivative

$$M^r_X(t) = E(X^r e^{tX}), \quad r \geq 1, \qquad (2.18)$$

and then for $t = 0$, we get

$$M^r_X(0) = E(X^r), \quad r \geq 1. \qquad (2.19)$$

The basic idea is that $M^r_X(0)$ equals the rth moment about the origin, that is, μ'_r.

Cumulant-Generating Functions

A cumulant-generating function is the log of the moment-generating function. Let us denote it by $K_X(t)$. This function can be written as an infinite series of the form

$$K_X(t) = \kappa_1 t + \kappa_2 \frac{t^2}{2} + \cdots + \kappa_r \frac{t^r}{r} + \cdots, \qquad (2.20)$$

where κ_r is known as the rth cumulant.

Remark   A cumulant-generating function can derive the moments around the mean and can determine the distribution of a random variable X. Like the moment-generating function, it can be used to obtain moments of a distribution. Additionally, cumulants are easier to deal with for sums of independent random variables and differences between distributions.

Using the definition $K_X(t) = \log_e M_X(t)$, one can derive a very useful set of equalities. Let us expand the mgf to get

$$\log_e M_X(t) = \log\left(1 + \mu'_1 t + \mu'_2 \frac{t^2}{2} + \cdots + \mu'_r \frac{t^r}{r} + \cdots\right)$$

$$= \left(\mu'_1 t + \mu'_2 \frac{t^2}{2} + \cdots\right) - \frac{1}{2}\left(\mu'_1 t + \mu'_2 \frac{t^2}{2} + \cdots\right)^2 + \cdots. \quad (2.21)$$

Comparing the coefficients of various powers of t for the cumulant- and moment-generating functions, one can easily derive the following equalities:

$$\kappa_2 = \mu'_2 - \mu'^2_1 = \mu_2,$$

$$\kappa_3 = \mu_3' - 3\mu_2'\mu_1' + 2\mu_1'^3 = \mu_3,$$
$$\kappa_4 = \mu_4' - 4\mu_3'\mu_1' + 6\mu_2'\mu_1'^2 - 3\mu_1'^4 - 3(\mu_2' - \mu_1'^2)^2 = \mu_4 - 3\kappa_2^2, \quad (2.22)$$

and so forth.

Characteristic Function

Sometimes the moment- and cumulant-generating functions do not exist due to the lack of convergence induced by real values of t. For a more general formulation, we can define a characteristic function with an imaginary or complex number i, such that $i^2 = -1$.

Let $\phi_X(t)$ be the characteristic function defined as

$$\phi_X(t) = E(e^{itX}) = \begin{cases} \sum_x e^{itx} g_X(x) & \text{for discrete variables,} \\ \int_{-\infty}^{\infty} e^{itx} f_X(x) dx & \text{for continuous variables.} \end{cases} \quad (2.23)$$

Remark Expanding the series of characteristic function, we get the moments as

$$\phi_X(t) = E(e^{itX}) = 1 + it\mu_1' + \frac{(it)^2}{2}\mu_2' + \cdots = \sum_{r=0}^{\infty} \frac{(it)^r}{r} \mu_r'. \quad (2.24)$$

We can also write the cumulant-generating function in terms of the characteristic function as

$$K_X(t) = \log_e \phi_X(t) = \sum_{r=0}^{\infty} \frac{(it)^r}{r} \kappa_r. \quad (2.25)$$

Example 2.6 Let X be a random variable with pdf $f_X(x) = 1, 0 < x < 1$. We write the mgf $M_X(t)$ as:

$$E\left(e^{tX}\right) = \int_0^1 e^{tX} dx = \frac{e^t - 1}{t} \quad \forall t \in \mathbb{R}.$$

Similarly, we write the characteristic function $\phi_X(t)$ as

$$E\left(e^{itX}\right) = \int_0^1 e^{itX} dx = \frac{e^{it} - 1}{it} \quad \forall t \in \mathbb{R}.$$

Using Euler's equation $e^{it} = \cos t + i \sin t$, we write

$$E\left(e^{itX}\right) = \frac{(\cos t + i \sin t - 1)}{it} = \frac{i(\cos t + i \sin t - 1)}{i^2 t}$$
$$= \frac{1}{t}(i - i\cos t - i^2 \sin t) = \frac{1}{t}(i - i\cos t + \sin t). \quad (2.26)$$

2.2.4 Discrete Distributions

Let us now consider X to be a discrete variable. Examples of such variables are from the number of cars passing by a road in an hour or the individual ages of people in a group. Some well-known distributions of discrete variables are: Bernoulli/binomial, Poisson, geometric, hypergeometric, negative binomial, etc. In the following sections we discuss the Bernoulli/binomial and Poisson distributions, which we will further discuss in analyzing complex system models in this book.

Bernoulli/Binomial Distribution

The Bernoulli distribution of a random variable X with parameter p is described as

$$g_X(x) = P(X = x) = \begin{cases} p^x(1-p)^{1-x} & \text{for } x = 0, 1, \\ 0 & \text{otherwise.} \end{cases} \tag{2.27}$$

Here the restriction is that p is between 0 and 1, that is, $0 \le p \le 1$. The outcome can be thought of to represent success ($x = 1$) or failure ($x = 0$). If we replicate an experiment n times with the outcomes $X_1, \ldots, X_n$ being independent Bernoulli(p), then the sum of those outcomes $Y = X_1 + \cdots + X_n$ will be binomial(n, p) random variable with the pmf

$$g_Y(y) = P(y = y) = \begin{cases} \binom{n}{y}p^y(1-p)^{n-y} & \text{for } x = 0, 1, \ldots, n, \\ 0 & \text{otherwise.} \end{cases} \tag{2.28}$$

Both n and p are parameters of the distribution. The notation $\binom{n}{y}$ represents the number of different ways an unordered set of y entities can be chosen from a set of n entities.

Remark   The expectation or mean of Bernoulli variable $E(X) = p$. For binomial distribution, it is $E(y) = np$ and the corresponding variances are $Var(X) = p(1-p)$ and $Var(y) = np(1-p)$, respectively.

Poisson Distribution

A random variable X following a Poisson distribution with parameter λ is described as

$$g_X(x) = P(X = x) = \begin{cases} \frac{\exp(-\lambda)\lambda^x}{x} & \text{for } x = 0, 1, \ldots; \lambda > 0, \\ 0 & \text{otherwise.} \end{cases} \tag{2.29}$$

Remark   The mean and variance of Poisson distribution are $E(X) = \lambda$ and $Var(X) = \lambda$.

Remark   It can be shown that the binomial distribution converges to Poisson distribution when the number of binomial trials are large, that is, $n \to \infty$, and the probability of success of the binomial trial is very small, that is, $p \rightarrow 0$, such that the product $np = \lambda$ where λ is a positive real number. This λ also turns out to be the mean of the Poisson distribution. This is a trick that we will use in the Kolkata paise restaurant problem described in Section 6.6.1.

Example 2.7 Suppose a coin is tossed 5 times and we know that the coin is slightly biased toward heads, with the probability of heads being 0.6. Suppose we are interested in finding the probability of getting at least 4 heads in 5 tosses. The probability of getting x number of heads in 5 independent tosses is given by

$$P(X = x) = \binom{5}{x} 0.6^x (1 - 0.6)^{5-x}. \tag{2.30}$$

Hence, the probability of getting at least 4 heads in 5 tosses is given by

$$P(X \geq 4) = P(X = 4) + P(X = 5)$$
$$= \binom{5}{4} 0.6^4 (1 - 0.6)^{5-4} + \binom{5}{5} 0.6^5 (1 - 0.6)^{5-5}$$
$$\approx 0.2592 + 0.0778 = 0.337. \tag{2.31}$$

Now suppose that we have 10,000 coins and we know that 1% of the coins are biased. We randomly select 1,000 coins from the pile of 10,000 coins and state that not more than 5 coins are defective. What is the probability that we are wrong in our statement? Clearly, here $n = 1,000$ and the probability that a coin is biased $p = 1\% = 0.01$. As n is relatively large and p is small, we can directly use the Poisson approximation of binomial distribution. The mean of the Poisson distribution $\lambda = np = 1,000 \times 0.01 = 10$. Using this value, we can write the probability of having x number of biased coins in a set of 1,000 coins as

$$P(X = x) = \frac{e^{-\lambda} \lambda^x}{x} = \frac{e^{-10} 10^x}{x}. \tag{2.32}$$

Therefore, we can calculate the probability that not more than 5 coins will be biased as

$$P(X \leq 5) = \sum_{x=0}^{5} \frac{e^{-10} 10^x}{x}. \tag{2.33}$$

Hence, the probability that we are wrong about the statement is

$$P(X > 5) = 1 - P(X \leq 5) = 1 - \sum_{x=0}^{5} \frac{e^{-10} 10^x}{x} \approx 0.9329. \tag{2.34}$$

2.2.5 Continuous Distributions

In this section, we will describe a few continuous distributions: normal, lognormal, exponential, gamma, and uniform distributions, among others. We will also describe extreme value distributions later in this section.

Normal Distribution

Normal distribution has two parameters: mean $\mu \in \mathbb{R}$ and variance $\sigma^2 > 0$. A normally distributed random variable X is written as $X \sim N(\mu, \sigma^2)$. The probability density function is written as

$$f_X(x) = \frac{1}{\sqrt{2\pi\sigma^2}} \exp\left(-\frac{1}{2\sigma^2}(x-\mu)^2\right) \quad \text{for } x \in \mathbb{R}. \tag{2.35}$$

The first two moments are given by $E(X) = \mu$ and $Var(X) = \sigma^2$, respectively.

Remark Let $Z = \frac{X-\mu}{\sigma}$. This scaled variable Z is said to follow a standard normal distribution with $E(Z) = 0$ and $Var(Z) = 1$. Notationally, $Z \sim N(0, 1)$. We can write the pdf of standard normal distribution as

$$f_Z(z) = \frac{1}{\sqrt{2\pi}} \exp\left(-\frac{1}{2}(z)^2\right) \quad \text{for } z \in \mathbb{R}. \tag{2.36}$$

A popular form of standard normal distribution is its distribution function $\Phi(z)$. We can write it in an integral form

$$\Phi(z) = P(Z \leq z) = \int_{-\infty}^{z} \left[\frac{1}{\sqrt{2\pi}} \exp\left(-\frac{1}{2}(u)^2\right)\right] du. \tag{2.37}$$

It can be shown easily that $\Phi(-z) = 1 - \Phi(z)$. Also, it follows that

$$P(a \leq X \leq b) = \Phi\left(\frac{b-\mu}{\sigma}\right) - \Phi\left(\frac{a-\mu}{\sigma}\right) \tag{2.38}$$

when $X \sim N(\mu, \sigma^2)$.

Remark Let $Y \in \mathbb{R}^+$ be a random variable. It follows a lognormal distribution if $\log_e(y)$ is normally distributed, that is, $\log_e(y) \sim N(\mu, \sigma^2)$. The lognormal distribution is written as

$$f_Y(y) = \begin{cases} \frac{1}{y\sqrt{2\pi\sigma^2}} \exp\left(-\frac{1}{2\sigma^2}(\log_e y - \mu)^2\right) & \text{for } y \geq 0, \\ 0 & \text{otherwise.} \end{cases} \tag{2.39}$$

Remark Let $X \sim N(\mu, \sigma^2)$ be a normally distributed variable and $Z = \frac{X-\mu}{\sigma}$ be a standard normal variable. A square transformation of it, $Z^2 = \left(\frac{X-\mu}{\sigma}\right)^2$, is a χ^2 (chi-square) variable with one degree of freedom. For n independent normal variates with mean μ_i and standard deviation σ_i, we can generalize it to

$$\chi^2 = \sum_{i=1}^{n} \left(\frac{X - \mu_i}{\sigma_i} \right)^2 \tag{2.40}$$

with n degrees of freedom. In such cases, we write that Y has a chi-square distribution with parameter (or degrees of freedom) n, that is, $Y \sim \chi^2_{(n)}$ with pdf

$$f_Y(y) = \begin{cases} \frac{1}{2^{n/2}\Gamma(n/2)} y^{(n/2)-1} \exp(-y/2) & \text{for } y \geq 0, \\ 0 & \text{otherwise.} \end{cases} \tag{2.41}$$

Example 2.8 Suppose from a dataset we find that the average temperature during July is 23.5°C with a standard deviation of 4°C. Let us say we select randomly a day in July and we are interested in finding the probability that the temperature will be between 18°C and 20°C. Under the assumption of normal distribution for temperature, the corresponding probability for X as the average temperature in °C would be

$$P(18 \leq X \leq 20) = P\left(\frac{18 - 23.5}{4} \leq \frac{X - 23.5}{4} \leq \frac{20 - 23.5}{4} \right)$$

$$= P\left(-1.375 \leq \frac{X - 23.5}{4} \leq -0.875 \right)$$

$$= \Phi(-0.875) - \Phi(-1.375)$$

$$= (1 - \Phi(0.875)) - (1 - \Phi(1.375))$$

$$\approx 0.191 - 0.085 = 0.106. \tag{2.42}$$

Exponential Distribution

Let X be a random variable with an exponential distribution characterized by parameter θ. We write its pdf as

$$f_X(x) = \begin{cases} \theta \exp(-\theta x) & \text{for } x \geq 0, \\ 0 & \text{otherwise.} \end{cases} \tag{2.43}$$

Remark The mean of exponential distribution is given by $E(X) = \frac{1}{\theta}$ and the variance is given by $Var(X) = \frac{1}{\theta^2}$.

Remark For a variable X with exponential distribution, one can show that

$$P(X \leq x + \Delta x | X \geq x) = P(0 \leq X \leq \Delta x).$$

In fact, one can show that an exponentially distributed variable X would have memoryless property:

$$P(X > x + \Delta x | X > x) = P(X > \Delta x), \quad x > 0. \tag{2.44}$$

Therefore, if we model a random event which happens with exponentially distributed time lags, the probability of an event within some Δx time is independent of when the last event took place.

Example 2.9 Suppose a research study of an RNA virus indicates that the time required for the virus to mutate is exponentially distributed. Let us assume that it takes on average 9 weeks (i.e., 63 days) for the virus to mutate into a modified RNA strain. Suppose we have a vaccine that works on the old strain. What is the probability that the vaccine will work if we start vaccination after week 5 of the first infection? We can answer this using the concept of exponential distribution. Let X denote the number of weeks required for mutation. Given the form of exponential distribution, we get $E(X) = 9 = \frac{1}{\theta}$, that is, $\theta = 1/9$ and hence the desired probability is

$$P(X > 5) = 1 - P(X \leq 5)$$
$$= 1 - \left(1 - e^{-\frac{1}{9} \times 5}\right) \approx 0.5738. \tag{2.45}$$

To derive the second equality from the first equality, we have utilized the expression of the cumulative distribution of exponential distribution.

Gamma Distribution

The parameters of gamma distribution are a slope $\alpha \in \mathbb{R}^+$ and a scale $\beta \in \mathbb{R}^+$. A gamma-distributed random variable X has the following probability density function:

$$f_X(x) = \begin{cases} \frac{\beta^\alpha}{\Gamma(\alpha)} x^{\alpha-1} \exp(-\beta x) & \text{for } x \geq 0, \\ 0 & \text{otherwise.} \end{cases} \tag{2.46}$$

Remark   Gamma distribution has mean $E(X) = \frac{\alpha}{\beta}$ and variance $Var(X) = \frac{\alpha}{\beta^2}$.

Uniform Distribution

Consider a continuous random variable X uniformly distributed over an interval (a, b). The corresponding probability density function is described as

$$f_X(a, b) = \begin{cases} \frac{1}{b-a} & \text{if } a < x < b \\ 0 & \text{otherwise.} \end{cases} \tag{2.47}$$

Remark The mean of the uniform distribution $E(X) = \frac{1}{2}(a + b)$ and its variance $Var(X) = \frac{1}{12}(b - a)^2$.

Generalized Extreme Value Distribution

Let X be a random variable that follows normal distribution with location parameter $\mu \in \mathbb{R}$ (i.e., mean) and scale parameter $\sigma \in \mathbb{R}^+$ (i.e., standard deviation). We write the generalized extreme value (GEV) distribution with a shape parameter $\zeta \in \mathbb{R}$ as follows:

$$f_X(x) = \begin{cases} \frac{1}{\sigma} \left(1 + \zeta \frac{x-\mu}{\sigma}\right)^{-1-\frac{1}{\zeta}} \exp\left(-\left(1 + \zeta \frac{x-\mu}{\sigma}\right)^{-\frac{1}{\zeta}}\right) & \text{for } \zeta \neq 0, \\ \frac{1}{\sigma} \exp\left(-\frac{x-\mu}{\sigma} - \exp\left(-\frac{x-\mu}{\sigma}\right)\right) & \text{for } \zeta = 0. \end{cases} \tag{2.48}$$

Remark The value of the shape parameter ζ can lead to different forms of the GEV distribution.

- For $\zeta > 0$, we obtain Fréchet distribution.
- For $\zeta < 0$, we obtain Weibull distribution.
- For $\zeta = 0$, we obtain Gumbel distribution.

Fréchet distribution does not have a finite right endpoint. Weibull, on the other hand, is a short-tailed distribution with a finite right endpoint. Gumbel distribution behaves the same way as Fréchet, but its tail decay faster.

Gumbel distribution is also known as the extreme value type-I distribution. For a random variable X, the probability density function of the Gumbel distribution is described as

$$f_X(x) = \frac{1}{\sigma} \exp\left(-\frac{x - \mu}{\sigma} - \exp\left(-\frac{x - \mu}{\sigma}\right)\right). \tag{2.49}$$

The Fréchet distribution is known as extreme value type-II distribution, with this probability density function:

$$f_X(x) = \frac{\zeta}{\sigma} \left(\frac{x - \mu}{\sigma}\right)^{-(1+\zeta)} \exp\left(-\frac{x - \mu}{\sigma}\right)^{-\zeta} \tag{2.50}$$

where the shape parameter $\zeta > 0$. Weibull distribution is known as extreme value type-III distribution, with this probability density function:

$$f_X(x) = \frac{\zeta}{\sigma} \left(\frac{x - \mu}{\sigma}\right)^{\zeta-1} \exp\left(-\frac{x - \mu}{\sigma}\right)^{\zeta} \tag{2.51}$$

where the shape parameter $\zeta < 0$.

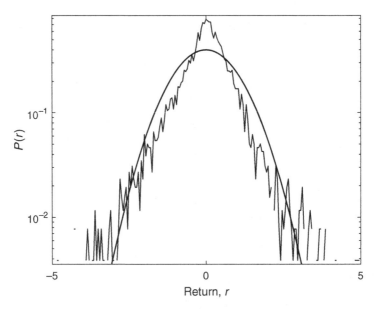

Figure 2.5 Distribution of (normalized) return on the S&P 500 index with a fitted normal distribution. The *y*-axis is shown in log scale. The return distribution is more peaked and wider than a normal distribution drawn with a solid curve.

Generalized Pareto Distribution

Let X be a normally distributed random variable with $\mu \in \mathbb{R}$ and $\sigma \in \mathbb{R}^+$ as the location and scale parameters. We write the generalized Pareto distribution (GPD) with a shape parameter $\zeta \in \mathbb{R}$ as

$$
f_X(x) = \begin{cases} \frac{1}{\sigma}\left(1 + \zeta\left(\frac{x-\mu}{\sigma}\right)\right)^{-1-\frac{1}{\zeta}}, & \zeta \neq 0, \\ & \text{for } \mu < x \text{ when } \zeta > 0 \text{ or} \\ & \text{for } \mu < x < \mu - \frac{\sigma}{\zeta} \text{ when } \zeta < 0, \\ 1 - \exp\left(-\left(\frac{x-\mu}{\sigma}\right)\right), & \zeta = 0 \text{ for } \mu < x. \end{cases} \tag{2.52}
$$

Remark   For the shape parameter $\zeta > 0$ and location $\mu = \frac{\sigma}{\zeta}$, the generalized Pareto distribution becomes Pareto distribution. For shape parameter $\zeta = 0$ and location parameter $\mu = 0$, the generalized Pareto distribution boils down to exponential distribution.

Empirical Example of Stock Return Distribution

In Figure 2.5, we have plotted the return distribution of the Standard & Poor's (S&P) 500 index with daily log returns over a period of around 20 years. (These data are analyzed in more detail from a time series perspective in Chapter 3.) Each day the market closes with an index value p_t – where t denotes the day, with values ranging from 1 to 5,175 here (corresponding to all the trading days over a period of 20 years). For the current purpose, it suffices to note that the daily returns are constructed as the first difference of the price series: $r_t = \log p_t - \log p_{t-1}$. Then we transform it into the z-score, that is, $r_t^z = (r_t - \bar{r})/\sigma_r$, where $\bar{r}$ denotes in-sample average and σ_r denotes in-sample standard deviation. The distribution of r_t is known to exhibit Pareto distribution in the tails (Mantegna and Stanley, 2007), especially with higher frequency data. Here we have plotted it along with the best fit of normal distribution. Two features are clearly visible. First, the empirical distribution is more peaked than the fitted normal distribution, and second, the fitted distribution poorly models the fluctuation in the tails. One question remains unanswered here: how did we estimate the parameters of the normal distribution given the data? For estimation, we employ the maximum likelihood estimator described in Section 2.3.

2.2.6 Statistical Approximations

Law of Large Numbers

The law of large number essentially says that the average from a large number of draws from a given distribution is approximately the same as the mean of the underlying distribution with high probability. Let X be the random variable with well-defined expectation $E(X)$ and variance $Var(X)$. Let us also define the sample analogue of the first moment as $\bar{X}_n = \frac{1}{n} \sum_{i=1}^{n} X_i$. The weak law of large number (WLLN) requires that for $\delta > 0$,

$$P(|\bar{X}_n - E(X)| > \delta) \to 0 \quad \text{as } n \to \infty. \tag{2.53}$$

The strong law of large number (SLLN) ensures *almost sure convergence* to the mean, that is, $\bar{X}_n - E(X) \to 0$ with probability 1 as $n \to \infty$.

Central Limit Theorem

The central limit theorem is a cornerstone result in probability theory. Loosely speaking, it states that for variables belonging to a large class of well-behaved distributions, the distribution of the mean of those variables converges to a normal distribution around the true mean as the sample size increases. This theorem is a very powerful tool for asymptotic theory in statistics.

There are multiple versions of this theorem. Here we state one of the simplest versions. If $X_1, X_2, \ldots, X_n$ are independent and identically distributed random samples of size n from a distribution with finite mean μ and finite variance σ^2, then the scaled mean

$$\frac{\bar{X} - \mu}{\sigma/\sqrt{n}} = \frac{\sqrt{n}(\bar{X} - \mu)}{\sigma}$$

converges to the standard normal distribution as $n \to \infty$. Here, $\bar{X} = \frac{1}{n} \sum_{i=1}^{n} X_i$. In a more concise manner, one can write

$$\lim_{n \to \infty} P\left(\frac{\sqrt{n}(\bar{X} - \mu)}{\sigma} \leq x\right) = \Phi(x) \quad \text{for } x \in \mathbb{R}, \tag{2.54}$$

where $\Phi(x)$ is the standard normal cumulative density function. Note that we did not require the variable X to be normally distributed although the transformation converges to a normal distribution.

Chebyshev's Inequality

Chebyshev's inequality is an important tool for obtaining probability bounds. Let X be a random variable with a distribution with finite mean μ and finite non-zero variance σ^2. For $c \geq 1$, we have

$$P\left(|X - \mu| \geq c \times \sigma\right) \leq \frac{1}{c^2}. \tag{2.55}$$

This inequality shows that the probability that X differs from mean μ by at least c times the standard deviation is less than or equal to $1/c^2$. We can rewrite the inequality as

$$P\left(|X - \mu| < c \times \sigma\right) \geq 1 - \frac{1}{c^2}. \tag{2.56}$$

While technically the above holds true for all $c > 0$, the range of c between 0 and 1 is not informative. The reason is that in such cases, the right-hand side becomes negative and probability has to be a positive number by definition. So the inequality would hold trivially. A similar argument in reverse would work for Equation 2.55 as well. For $c < 1$, the right-hand side would be more than 1 and probability has to be less than 1, again by definition, implying that the inequality would hold trivially.

2.2.7 Sampling Distribution

Sampling distribution refers to the distribution of a *statistic* obtained from a repeated sampling scheme from a population. Here it is useful to recall from our earlier discussion that a *statistic* is a function of the underlying observations from the sample. For example, let $X_1, X_2, \ldots, X_n$ be the sample observations drawn from a population with normal distribution $N(\mu, \sigma^2)$. The corresponding sample mean is simply $\bar{X} = \frac{1}{n} \sum_{i=1}^{n} X_i$. This statistic has a distribution $N(\mu, \sigma^2/n)$. If $s^2 = \frac{1}{n-1} \sum_{i=1}^{n} (X_i - \bar{X})^2$ is the corresponding sample variance, then $\frac{(n-1)s^2}{\sigma^2}$ follows a χ^2 distribution with $n - 1$ degrees of freedom, the variables $\bar{X}$ and s^2 are independent.

Standard error is a statistic that measures how well a *sampling distribution* captures the *population distribution* in terms of standard deviation. The standard error of the mean measures the deviation of the sample mean from the population mean.

Using the same notation, let us consider $X_1, X_2, \ldots, X_n$ to be the observation of n random sample drawn from a population with distribution $N(\mu, \sigma^2)$. The standard deviation is $s = \sqrt{\frac{1}{n-1} \sum_{i=1}^{n} (X_i - \bar{X})^2}$. The standard error of the mean is defined as $se(\bar{X}) = \frac{s}{\sqrt{n}}$. For calculating the sample mean, an increase in the number of observations n clearly yields a smaller standard error.

2.3 Classical Statistical Inference

One of the key objectives in statistics is to make inferences about a population using information obtained from a sample by utilizing the observed probability distributions (Hogg et al., 2005). We know that population distribution is characterized by numerical descriptive measures: that is, *parameters*. Our objective in statistical inference is to estimate the population parameters. We call the corresponding measure a *statistic*. Simply put, a parameter relates to the population and a statistic represents the corresponding measurement from samples drawn from the population.

2.3.1 Estimation

Estimation is a procedure by which we can gauge the true value of a population parameter, which is unknown to us *a priori*, by using sample observations (Rice, 2006). Let $X_1, X_2, \ldots, X_n$ be a random sample from a distribution $f(x, \theta)$ where θ is the unknown parameter. A statistic $T_n = t(X_1, \ldots, X_n)$ used to estimate $g(\theta)$, some function of θ, is an *estimator* of $g(\theta)$, and any value of $t(X_1, \ldots, X_n)$ is an *estimate* of $g(\theta)$.

An estimation procedure is divided into (i) point and (ii) interval estimation, where the sample information is used to estimate (i) a single number and (ii) an interval for the population parameter.

Properties of a Good Estimator

A good estimator has five properties: unbiasedness, consistency, efficiency, sufficiency, and completeness. Among these, the ones defined for increasing sample size are known as large sample properties, such as consistency, whereas the ones defined for a given sample size are referred to as small sample properties, such as unbiasedness.

Unbiasedness: Consider $T_n = t(X_1, \ldots, X_n)$, a statistic calculated from the sample. This estimator is defined to be an unbiased estimator of $g(\theta)$ if $E(T_n) = g(\theta)$. If $E(T_n) \neq g(\theta)$, then the statistic is said to be a biased estimator for $g(\theta)$.

Consistency: A statistic T_n is said to be a consistent estimator if, for any positive number δ, we have $\lim_{n \to \infty} P(|T_n - g(\theta)| \leq \delta) = 1$ or $\lim_{n \to \infty} P(|T_n - g(\theta)| > \delta) = 0$.

Efficiency: A statistic T_n is said to be an efficient estimator if T_n is unbiased and $Var(T_n) \leq Var(T'_n)$ where T'_n is any other unbiased estimator of $g(\theta)$.

Sufficiency: A statistic T_n is a sufficient estimator of $g(\theta)$ if, and only if, the conditional distribution of the sample observations $X_1, X_2, \ldots, X_n$ given T_n does not depend on $g(\theta)$ for any value of T_n; that is, once T_n is known, no other function of the sample observations will provide additional information on the possible values of $g(\theta)$. In such cases, we describe T_n to be sufficient and it contains all the information about $g(\theta)$ gathered from the samples.

Completeness: A statistic T_n is defined to be complete if, and only if, when $E(h(T_n)) = 0$ for all $\theta \in \Theta$ where Θ is the parameter space, then $P(h(T_n) = 0) = 1$ for all $\theta \in \Theta$, where $h(.)$ is a function of the statistic T_n.

Example 2.10 Let $X_1, X_2, \ldots, X_n$ be a random sample from normal distribution with mean μ and variance σ^2. We define the sample mean $\bar{X} = \frac{1}{n} \sum_{i=1}^{n} X_i$, which is an unbiased estimator of the population mean μ, as

$$E(\bar{X}) = E\left(\frac{1}{n} \sum_{i=1}^{n} X_i\right) = \frac{1}{n} \sum_{i=1}^{n} E(X_i) = \frac{1}{n} \sum_{i=1}^{n} \mu = \frac{1}{n} n\mu = \mu. \quad (2.57)$$

However, if we define sample variance analogous to population variance by dividing by n instead of the degrees of freedom $n - 1$, that is, $s_b^2 = \frac{1}{n} \sum_{i=1}^{n} (X_i - \bar{X})^2$, then s_b^2 becomes a biased estimator of the population variance σ^2. Note that

$$E(s_b^2) = E\left(\frac{1}{n} \sum_{i=1}^{n} (X_i - \bar{X})^2\right) = E\left(\frac{1}{n} \left(\sum_{i=1}^{n} X_i^2 - n\bar{X}^2\right)\right)$$

$$= E\left(\frac{1}{n} \sum_{i=1}^{n} X_i^2 - \bar{X}^2\right) = \frac{1}{n} \sum_{i=1}^{n} E(X_i^2) - E(\bar{X}^2). \quad (2.58)$$

We know $\sigma^2 = Var(X_i) = E(X_i^2) - E(X_i)^2$, thus $E(X_i^2) = \sigma^2 + \mu^2$ as $E(X_i) = \mu$. We also get $Var(\bar{X}) = E(\bar{X}^2) - E(\bar{X})^2$ and $Var(\bar{X}) = Var\left(\frac{1}{n} \sum_{i=1}^{n} X_i\right)$, that is, $Var(\bar{X}) = \sigma^2/n$. Hence, we get

$$E(s_b^2) = \frac{1}{n} \sum_{i=1}^{n} E(X_i^2) - E(\bar{X}^2) = \frac{1}{n} \sum_{i=1}^{n} E(X_i^2) - Var(\bar{X}) + E(\bar{X})^2$$

$$= \frac{1}{n} \sum_{i=1}^{n} (\sigma^2 + \mu^2) - \left(\frac{\sigma^2}{n} + \mu^2\right)$$

$$= \left(1 - \frac{1}{n}\right) \sigma^2. \quad (2.59)$$

This means that s_b^2 is not an unbiased estimator of σ^2. However, if we consider $s^2 = \frac{1}{n-1}\sum_{i=1}^{n}(X_i - \bar{X})^2$, then we get

$$E(s^2) = E\left(\frac{1}{n-1}\sum_{i=1}^{n}X_i\right) = \frac{n}{n-1}E\left(\frac{1}{n}\sum_{i=1}^{n}X_i\right)$$

$$= \frac{n}{n-1}\left(1 - \frac{1}{n}\right)\sigma^2 = \sigma^2. \tag{2.60}$$

Hence, keeping $n - 1$ in the denominator makes the estimator unbiased. Note that the sample variance s^2 is a consistent estimator of the population variance σ^2, that is, when sample size n is large, then s^2 converges to σ^2 with high probability.

Estimation with Method of Moments

To estimate unknown population parameters, we can use a method based on moments (see details in Section 2.2.3). This approach is relatively simple compared to other estimation techniques for parameters.

Let $X_1, X_2, \ldots, X_n$ be a random sample from a distribution $f(x, \theta)$ with one-parameter θ, and let $E(X) = \mu' = \mu'(\theta)$ be the first population moment. The method of moments consists of solving the equation for θ in terms of μ by replacing the population moment by its sample counterpart $\hat{\mu}$, that is, $\hat{\theta} = \theta(\hat{\mu})$. The same method can be used for higher-order moments as well.

The method of moment estimator can only be applied to estimate a given number of parameters in the population distribution, as per the moment conditions. For example, let X be a normally distributed random variable with two parameters μ and σ^2 denoting mean and variance, respectively. The method of moments is then applied only for estimating the mean and variance of the normal distribution from the sample moments with exact solutions.

Estimation with Generalized Method of Moments

As the name suggests, this technique is a generalization of the method of moment estimators (Hansen, 1982). Let us consider a case where the number of moments r exceeds the number of parameters k in the population distribution. In such cases, we do not get exact solutions because we have k unknowns and r equations with $r > k$. To solve such problems, we have to define some cost functions such that minimizing the cost functions produces a solution.

Let $X_1, X_2, \ldots, X_n$ be a random sample from $f(x, \theta)$ with one-parameter θ, that is, $k = 1$; and let $E(x^r) = \mu_r' = \mu_r'(\theta)$ be the rth population moment where $r > 1$. We define the rth cost function $g_r(\theta)$ and minimize $g_r(\theta)$ with respect to θ to estimate $\hat{\theta}$.

Method of moment (or generalized method of moment) estimators are generally not efficient, but under suitable conditions they can be consistent. Furthermore, they may not be functions of sufficient or complete statistics.

Example 2.11   Suppose $X_1, X_2, \ldots, X_n$ is a random sample from a normal distribution with mean μ and variance σ^2, that is, $X \sim N(\mu, \sigma^2)$. We write the first moment about the origin as

$$E(X) = \mu'_1 = \frac{1}{n} \sum_{i=1}^{n} X_i. \tag{2.61}$$

The method of moment estimator for the mean μ is the sample mean

$$\hat{\mu}_1 = \frac{1}{n} \sum_{i=1}^{n} X_i = \bar{X}. \tag{2.62}$$

Now, the second moment about the origin is

$$E(X^2) = \mu'_2 = \frac{1}{n} \sum_{i=1}^{n} X_i^2 \tag{2.63}$$

and the second moment about the mean is given by

$$\hat{\mu}_2 = \mu'_2 - \mu'^2_1 = \frac{1}{n} \sum_{i=1}^{n} X_i^2 - \bar{X}^2. \tag{2.64}$$

We write the method of moment estimator for the variance σ^2 as

$$\hat{\sigma}^2 = \frac{1}{n} \sum_{i=1}^{n} X_i^2 - \bar{X}^2$$

$$= \frac{1}{n} \sum_{i=1}^{n} X_i^2 - 2\bar{X}^2 + \bar{X}^2$$

$$= \frac{1}{n} \sum_{i=1}^{n} \left(X_i^2 - 2X_i\bar{X} + \bar{X}^2 \right)$$

$$= \frac{1}{n} \sum_{i=1}^{n} (X_i - \bar{X})^2. \tag{2.65}$$

Estimation Using Likelihood Function

Maximum likelihood is a very commonly used method for estimation. In this method, one constructs likelihood functions from the observed data in conjunction with distributional assumptions. The goal of the modeler is to maximize this likelihood function to estimate the population parameters. As the name suggests, this process leads to the maximum likelihood estimates of the parameters.

Let $X_1, X_2, \ldots, X_n$ be a random sample from a distribution $f(x, \theta)$. Let us denote the parameter space by Θ. The likelihood function $L(\theta) = L(\theta; X)$ is defined by

$$L(\theta) = \prod_{i=1}^{n} f(x_i, \theta). \tag{2.66}$$

If $\hat{\theta}$ is an estimated value of $\theta \in \Theta$ which maximizes the likelihood function $L(\theta)$, then $\hat{\theta}$ is known as the *maximum likelihood estimator* (MLE) of θ.

If a likelihood function contains p number of parameters $(\theta_1, \ldots, \theta_p)$, then we denote the likelihood function $L(\theta_1, \ldots, \theta_p) = \prod_{i=1}^{n} f(x_i, \theta_1, \ldots, \theta_p)$. We want to maximize the likelihood function with respect to its arguments. Therefore, *joint likelihood estimators* can be obtained by solving the following p equations:

$$\frac{\partial}{\partial \theta_i} \log L(\theta_1, \ldots, \theta_p) = 0; \quad i = 1, \ldots, p \tag{2.67}$$

with a sufficient condition that the matrix $\left[\frac{\partial^2}{\partial \theta_i \partial \theta_j} \log L(\theta_1, \ldots, \theta_p) \right]_{p \times p}$ is negative definite. An MLE is not necessarily unique. If it exists, it is a function of a sufficient statistic for the population parameters. Furthermore, an MLE may not be unbiased for samples. However, if $\hat{\theta}$ is the MLE of θ, then $h(\hat{\theta})$ is the MLE of $h(\theta) \in \Theta$.

Technically speaking, to show that the MLE is maximum, we need to show that the *Hessian* matrix is negative definite; that is,

$$[\mathbf{H}(\theta)] = \frac{\partial^2}{\partial \theta_i \partial \theta_j} \log L(\theta)$$

is negative definite at $\theta = \hat{\theta}$, that is, $[\mathbf{H}(\hat{\theta})] < 0$.

Remark We can find *Fisher's information* using the Hessian matrix

$$[I(\theta)] = E(-[\mathbf{H}(\theta)]).$$

In likelihood theory, the variance–covariance matrix of the *score* vector can be written as the expected *information* matrix $Var(u(\theta)) = [I(\theta)]$ where we define the *score* function as

$$u_i(\theta) \equiv \frac{\partial}{\partial \theta_i} L(\theta); \quad i = 1, \ldots, p. \tag{2.68}$$

Note that $E(u(\theta)) = 0$ and the MLE of θ, that is, $\hat{\theta}$ satisfies $u(\hat{\theta}) = 0$.

Example 2.12   Let $X \sim N(\mu, \sigma^2)$. We can write the likelihood function as

$$L(\mu, \sigma^2 | X) = \prod_{i=1}^{n} \frac{1}{\sqrt{2\pi\sigma^2}} \exp\left(-\frac{1}{2\sigma^2}(x_i - \mu)^2\right)$$

$$= \left(\frac{1}{2\pi\sigma^2}\right)^{-n/2} \exp\left(-\frac{1}{2\sigma^2} \sum_{i=1}^{n}(x_i - \mu)^2\right). \tag{2.69}$$

For finding the estimate of μ, we take the derivative of this function with respect to μ and equate it to zero. To write the score function, we first take a logarithmic transformation for ease in computation of the derivatives. The resulting expression is

$$\log L(\mu, \sigma^2 | X) = -\frac{n}{2}\left(\log 2\pi + \log \sigma^2\right) - \frac{1}{2\sigma^2} \sum_{i=1}^{n}(x_i - \mu)^2$$

$$\frac{\partial}{\partial \mu} \log L(\mu, \sigma^2 | X) = \frac{1}{2\sigma^2} \sum_{i=1}^{n}(x_i - \mu) = \frac{n}{\sigma^2}(\bar{x} - \mu). \tag{2.70}$$

Assuming that the second derivative with respect to μ is negative, $\hat{\mu} = \bar{x}$ is the maximum likelihood estimator of μ.

Similarly, we can get the maximum likelihood estimate for σ^2:

$$\frac{\partial}{\partial \sigma^2} \log L(\mu, \sigma^2 | X) = -\frac{n}{2\sigma^2} + \frac{1}{2(\sigma^2)^2} \sum_{i=1}^{n}(x_i - \mu). \tag{2.71}$$

By substituting $\hat{\mu} = \bar{X}$, we write $\hat{\sigma}^2 = \frac{1}{n}\sum_{i=1}^{n}(X_i - \bar{X})^2$. Note that the maximum likelihood estimate $\hat{\sigma}^2$ is a biased estimator for σ^2. If we want unbiasedness, we can go for the estimator of σ^2 as $s^2 = \frac{1}{n-1}\sum_{i=1}^{n}(X_i - \bar{X})^2$.

Empirical Example: Household Consumption Distribution

To demonstrate an example of point estimation of distributional parameters, in Figure 2.6 we plot consumption distribution across households in India, along with fitted gamma and lognormal distributions. The household-level consumption data are obtained from the Consumer Pyramids Household Surveys (conducted by the Centre for Monitoring Indian Economy). We have used maximum likelihood estimators to fit the distribution. The exact algebraic form of the consumption distribution (even more so for income and wealth) has been a source of controversy in the economics literature. While the bulk seems to have a good fit with both gamma and lognormal distributions, a bigger discrepancy arises when modeling

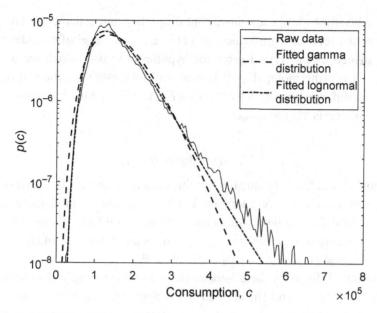

Figure 2.6 Empirical histogram of consumption distribution from a sample of ~83,000 households sampled across India in 2019. We show the data in log scale which fits with gamma distribution and lognormal distribution. Gamma distribution captures the bulk of the distribution, and lognormal distribution captures the tail better. We will analyze the corresponding income distribution later in Chapter 6 (see Figure 6.3).

the right tail of the distribution as both power law and lognormal distributions are good candidates for modeling the tail (see e.g. Chakrabarti et al. [2018]). There are multiple models that generate lognormal and power law distributions (see e.g. Mitzenmacher [2004]). We will pick up this thread of discussion again in Chapter 6 when we talk about kinetic exchange models.

Interval Estimation

In interval estimation, we define an interval which has a high probability of including the estimated population parameters. A confidence interval indicates that we have some measure of assurance or confidence that the population parameter lies within the interval.

Let $X_1, X_2, \ldots, X_n$ be a random sample from $f(x, \theta)$ and $T_{\text{low}} = t_{\text{low}}(X_1, \ldots, X_n)$ be a statistic for which $P(T_{\text{low}} \leq \theta) \geq 1 - \alpha$. T_{low} is known as the *lower* bound for θ with confidence level $(1 - \alpha)$. Similarly, if $T_{\text{up}} = t_{\text{up}}(X_1, \ldots, X_n)$ satisfies $P(\theta \leq T_{\text{up}}) \geq 1 - \alpha$, then T_{up} is called the *upper* bound for θ with confidence level $(1 - \alpha)$. Now suppose the two statistics satisfy $T_{\text{low}} \leq T_{\text{up}}$ for which $P(T_{\text{low}} \leq \theta \leq T_{\text{up}}) \geq 1 - \alpha$, then the interval $[T_{\text{low}}, T_{\text{up}}]$ is called a *two-sided* $(1 - \alpha)$ *confidence interval* for θ.

Here we briefly review some important properties and statements. The term α is known as the level of significance, and $(1 - \alpha)$ is the level of confidence. Note that α is treated as the Type I error for hypothesis testing which we discuss in Section 2.3.2. Furthermore, if $g(\theta)$ is a strictly increasing function of θ, then a two-sided confidence interval for $g(\theta)$ is $[g(T_{\text{low}}), g(T_{\text{up}})]$. Finally, we note that the confidence intervals are not unique.

2.3.2 Hypothesis Testing

In hypothesis testing, we statistically test the validity of an educated guess or statement about the data under consideration. Let $X_1, \ldots, X_n$ be the random sample from a population distribution with parameter θ. A *simple null hypothesis* is defined as $H_0 : \theta = \theta_0$, whereas the *alternative hypothesis* is a complement of H_0, which can be either $H_1 : \theta \neq \theta_0$ or $H_1 : \theta < \theta_0$ or $H_1 : \theta > \theta_0$.

Two possible inferences for a hypothesis test are: (i) sample data support H_1, that is, H_0 is rejected; and (ii) data do not support H_1 exclusively, that is, H_0 is not rejected. The first inference is more conclusive, that is, there is no evidence based on the available data that H_1 is incorrect. In contrast, the second inference is indecisive, that is, the sample data fail to reject H_0. A conservative interpretation of hypothesis testing is that H_1 may be correct when H_0 is rejected; however, it does not imply acceptance of H_0 when H_0 is not rejected. Generally speaking, hypothesis testing is much more geared toward rejection (or not) of H_0 rather than supporting (or not) H_1. One can conduct statistical test depending on the state of the alternative hypothesis. For example, for $H_1 : \theta \neq \theta_0$, we use a two-tailed test. For $H_1 : \theta < \theta_0$ or $H_1 : \theta > \theta_0$, a one-tailed test is used.

Remark A *composite hypothesis* is specified in a way which does not distinguish between multiple possibilities. A simple example is $H_0 : \theta > \theta_0$. In this case, the parameter θ has more than one possible value. The test would be inconclusive about the precise value of the parameter.

Type I and Type II Errors and Statistical Power

Under the classical approach, hypothesis testing is mainly dominated by the Neyman–Pearson paradigm where decision problem is defined under Types I and II errors (Rice, 2006). However, the basic intuition is applicable beyond statistics. As we will see in Chapter 4, a similar idea of model evaluation will come into play.

Let C denote the critical or rejection region for some statistic x under consideration. A hypothesis test can produce the following errors (see Table 2.1):

- Type I error: We reject H_0 when it is true, that is, $\alpha = P(\text{Type I error}) = P(x \in C | H_0 \text{ is true})$.

Table 2.1 *Decision rule*

		Sample	
		Reject H_0	Fail to reject H_0
Population	True H_0	Type I error	Correct
	False H_0	Correct	Type II error

- Type II error: We fail to reject H_0 when it is false, that is, $\beta = P(\text{Type II error}) = P(x \notin C | H_0 \text{ is not true})$.

Remark The probability of Type I error is known as the *size* of the test with its upper bound being called *significance level*. On the other hand, $\omega \equiv 1 - \beta$ is called the *power* of the test. In an ideal scenario, we would like to avoid both types of errors, that is, we would expect small values for α and β. From the modeler's perspective, a good test should have small *significance level* and large *power*. Additionally, we note that Type I error (α) can be more problematic than Type II error (β) as β leads to the indecisive inference of not rejecting H_0. Hence, the classical approach is to set up a control over α with a small value (e.g. 0.01 or 0.05) to limit the error. Unfortunately, we cannot simultaneously reduce the probabilities of both types of errors in practice. If we want to reduce the probability of Type I error, then usually the probability of Type II error increases, and vice versa.

Neyman–Pearson Lemma

This is an important result that allows us to identify the best hypothesis with the largest statistical power ($\omega = 1 - \beta$) while maintaining a small probability of Type I error.

Among all tests of significance level less than or equal to α of $H_0 : \theta = \theta_0$ against $H_1 : \theta = \theta_1$, the test that minimizes the probability of Type II error (i.e., largest power $\omega = (1 - \beta)$) is given by the likelihood-ratio test.

Suppose we want to test $H_0 : \theta = \theta_0$ against $H_1 : \theta \neq \theta_0$. Under the null hypothesis, we define the parameter space Θ_0. A likelihood ratio is defined as

$$\Lambda = \frac{L(\Theta_0)}{L(\Theta)} = \frac{L(x|\theta_0)}{L(x|\hat{\theta})} \tag{2.72}$$

where $\hat{\theta}$ is the maximum-likelihood estimate of θ (we assume that MLE exists). The test based on Λ is known as the likelihood-ratio test with a critical region

$$C = \left\{ x : \frac{L(x|\theta_0)}{L(x|\hat{\theta})} > k \right\} \tag{2.73}$$

where k is determined by the *size* or *significance* of the test

$$P(x \in C) = P\left(\frac{L(x|\theta_0)}{L(x|\hat{\theta})} > k\right) = \alpha. \tag{2.74}$$

Asymptotically, $-2\log_e \Lambda$ follows a chi-square distribution and the statistic is consistent.

2.3.3 Parametric Tests

In this section, we will briefly provide the mathematical definitions of some commonly used statistical parametric tests. We will also review the concept of p-value.

Z-Test

A Z-test is used to test a hypothesis where the test statistic can be assumed to follow a normal distribution, at least in the sense of an approximation. It can be used for a mean test or a test for proportions. See Table 2.2 for a summary.

Let $X_1, \ldots, X_n$ be the random sample from $f(x, \theta)$ where n is relatively large (at a minimum, there should be 30 data points; however, this number is merely indicative and application of this test depends on the type of experiment) and θ is the parameter under consideration. The null hypothesis could be that the sample estimate of θ is θ_0. We write the Z statistic for testing θ as

$$Z = \frac{\hat{\theta} - \theta_0}{se(\hat{\theta})} \tag{2.75}$$

where $\hat{\theta}$ is an estimate of θ (e.g. MLE) and $se(\hat{\theta}) = \sqrt{Var(\hat{\theta})/n}$ is the corresponding standard error of the estimate. We reject H_0 by comparing Z_{cal} with Z_{tab} where Z_{cal} is the Z-value calculated from the data and Z_{tab} can be obtained from the Z-table or generated by statistical software. As a natural cut-off, $Z_{tab} = 1.96$ for 5% level of significance.

T-Test

When the sample size is relatively small (a commonly used threshold is $n < 30$) and the goal is to compare the means of two samples, a t-test can be useful.

Let $X_1, \ldots, X_{n_x}$ and $Y_1, \ldots, Y_{n_y}$ be the random samples from $f(x, \theta_x)$ and $f(y, \theta_y)$. Let's say θ_x and θ_y represent the means of the distributions. We estimate the corresponding sample means $\bar{X}$ and $\bar{Y}$. The statistic is defined as

$$t = \frac{\bar{X} - \bar{Y}}{s_{x,y}^p \sqrt{1/n_x + 1/n_y}} \tag{2.76}$$

Table 2.2 Hypothesis tests and corresponding critical region

Hypothesis	Test statistic and critical region
$H_0 : \theta = \theta_0$, $H_1 : \theta > \theta_0$, σ^2 known	$Z = \frac{\bar{X} - \theta_0}{\sigma/\sqrt{n}} > Z_\alpha$
$H_0 : \theta = \theta_0$, $H_1 : \theta > \theta_0$, σ^2 unknown	$Z = \frac{\bar{X} - \theta_0}{s/\sqrt{n}} > t_\alpha(n-1)$
$H_0 : \sigma^2 = \sigma_0^2$, $H_1 : \sigma^2 > \sigma_0^2$, sample variance s^2	$\chi^2 = \frac{(n-1)s^2}{\sigma_0^2} > \chi_\alpha^2(n-1)$
$H_0 : \theta_x - \theta_y = 0$, $H_1 : \theta_x - \theta_y > 0$, $\sigma_x^2 = \sigma_y^2$ unknown	$t = \frac{\bar{X} - \bar{Y}}{s_{x,y}^p \sqrt{1/n_x + 1/n_y}} > t_\alpha(n_x + n_y - 2)$ where $s_{x,y}^p = \sqrt{\frac{(n_x-1)s_x^2 + (n_y-1)s_y^2}{n_x + n_y - 2}}$
$H_0 : \sigma_x^2/\sigma_y^2 = 1$, $H_1 : \sigma_x^2/\sigma_y^2 > 1$	$F = \frac{s_x^2}{s_y^2} > F_\alpha(n_x - 1, n_y - 1)$

where $s_{x,y}^p$ is the pooled standard deviation across both the samples; see Table 2.2 for a description. We reject H_0 if the t-value calculated from the sample $t_{cal} > t_{tab(n-1)}$ for $(n-1)$ degrees of freedom, where t_{tab} can be obtained from the t-table or generated by statistical software.

Chi-Square Test

For $X_1, X_2, \ldots, X_n$, n independent normal variable with mean μ_i and variance σ_i^2, we write the chi-square statistic with n degrees of freedom as

$$\chi^2 = \sum_{i=1}^n \left(\frac{X_i - \mu_i}{\sigma_i} \right)^2. \qquad (2.77)$$

H_0 is rejected if the statistic calculated from the sample $\chi_{cal}^2 > \chi_{tab(n-1)}^2$ for $n-1$ degrees of freedom. See Table 2.2 for a summary. χ_{tab}^2 can be obtained from a χ^2 table or generated by statistical software.

Remark A simplified version with one degree of freedom can be written as the square of a standard normal variate. If $X \sim N(\mu, \sigma^2)$, $Z = \frac{X-\mu}{\sigma} \sim N(0, 1)$. Thus, $Z^2 = \left(\frac{X-\mu}{\sigma}\right)^2$ has a chi-square distribution with one degree of freedom.

F-Test

Consider two independent chi-square variates U and V with degrees of freedom u and v, respectively. The F-statistic is the ratio of the scaled chi-square variates:

$$F = \frac{U/u}{V/v}. \qquad (2.78)$$

H_0 is rejected if the statistic calculated from the sample $F_{cal} > F_{tab(u,v)}$ for u and v degrees of freedoms. See Table 2.2 for a summary. The value of F_{tab} can be obtained from an F table or generated by statistical software.

ANOVA Test

ANalysis Of VAriance (shortened as ANOVA) is an extension of the two-sample t-test where the goal is to compare more than two variables or to compare multiple groups within a variable. Here we will consider two types of variation in the observed data: (i) within and (ii) between variables or groups. Let X_{ij} be the random sample of jth observation $j = 1, \ldots, n$ and the ith group $i = 1, \ldots, K$, where $n \times K$ is the total number of observations or samples. Considering equal number of samples from each group, that is, a balanced design, we write

$$X_{ij} = \bar{X} + (\bar{X}_i - \bar{X}) + (X_{ij} - \bar{X}_i)$$
$$= \bar{X} + \alpha_i + \epsilon_{ij} \tag{2.79}$$

where $\bar{X}$ is the grand sample mean, $\alpha_i = (\bar{X}_i - \bar{X})$ is the deviation of the sample mean for the ith group, $\epsilon_{ij} = (X_{ij} - \bar{X}_i)$ is the error which is assumed to follow a normal distribution – $\epsilon_{ij} \sim N(0, \sigma^2)$. To test the mean differences (α_i, for $i = 1, 2, \ldots, K$) we write the null hypothesis as $H_0 : \alpha_1 = \alpha_2 = \cdots = \alpha_K$. Under ANOVA, the goal is to decompose the total variation into the sum of squares between groups (labeled SSB) and sum of squares within groups (labeled SSW):

$$\text{SSB} = \sum_{i=1}^{K} \sum_{j=1}^{n} (\bar{X}_i - \bar{X})^2$$

$$\text{SSW} = \sum_{i=1}^{K} \sum_{j=1}^{n} (X_{ij} - \bar{X}_i)^2 \tag{2.80}$$

where SSB and SSW have $K - 1$ and $N - K$ degrees of freedom, respectively. This leads to the ANOVA test statistic being a ratio of SSB and SSW, which follows F-distribution with $K - 1$ and $N - K$ degrees of freedom, that is,

$$F = \frac{SSB/(K-1)}{SSW/(N-K)}. \tag{2.81}$$

The null hypothesis is that all group means are identical and the alternative hypothesis is that there is at least one pair for which equality in the means does not hold.

Concept of a p-Value

A p-value (or probability value) of a test is a particular probability associated with the test. Let us consider a null hypothesis $H_0 : \theta = \theta_0$ and corresponding alternative $H_0 : \theta > \theta_0$ where θ_0 is some specific value, say 13. Now, assume that we have a random sample from the population $f(x, \theta)$ and we have calculated the sample mean as $\bar{X} = 13.3$ from $n = 300$ samples with a known variance of 4. Here, the p-value represents the probability of obtaining $\bar{X}$ greater than or equal to 13.3 given the null hypothesis, that is, $P(\bar{X} \geq 13.3 | H_0)$. More generally, the p-value is the probability of getting a statistic at least as extreme as the observed one in the sample, conditional on the null hypothesis being actually true.

The smaller the p-value, the stronger becomes the evidence that a null hypothesis H_0 must be rejected. Following standard practice, let us consider $\alpha = 0.05$ to be the level of significance. Then a p-value less than 0.05 tells us that the test is statistically significant at $\alpha = 0.05$. Note that the p-value is only associated with the null hypothesis H_0; it does not provide any information about the alternative hypothesis H_1. For a more elaborate discussion, see Rice (2006) and Wood (2015) which provide a broader background for this concept.

2.3.4 Non-parametric Tests

When the population distribution is not known, we cannot go for parametric tests. The remedy in such cases is to consider non-parametric (also described as distribution-free) tests. One of the key advantages of such tests is that they exhibit good performance even for small samples.

Let $X_1, \ldots, X_n$ and $Y_1, \ldots, Y_m$ be the two sets of sample observations. Suppose we want to compare the distribution of X with Y. If n and m are large, then without making distribution assumptions about X and Y, we can utilize the *central limit theorem* which asymptotically gives us a normal distribution. Due to its ease of application, this theorem has become popular in parametric tests. However, when n and m are not large enough, then the theorem may not be applicable. In such cases, we can utilize the non-parametric methods to test the null hypothesis $H_0 : \mu_x = \mu_y$ against appropriate alternative $H_1 : \mu_x \neq \mu_y$ or $\mu_x < \mu_y$ or $\mu_x > \mu_y$ depending on the context. Many of these tests rely on rank-ordering observations and carrying out tests on the output. Here we briefly review a few such tests which are most commonly used. Interested readers can refer to Kvam and Vidakovic (2007) which discusses these tests and covers the background in more detail.

Mann–Whitney U Test

This non-parametric test can be applied for checking whether two samples are drawn from the same population or not. It is also known as the Wilcoxon rank-sum test. Here the null hypothesis is that the distributions of random variables X and Y are identical. Naturally, the alternative is that the distributions are not identical. We write the U statistic for X and Y as

$$U_X = n_x n_y + \frac{(1 + n_x) n_x}{2} - R_x,$$

$$U_Y = n_x n_y + \frac{(1 + n_y)n_y}{2} - R_y \qquad (2.82)$$

where n_x and n_y are sample sizes, and R_x and R_y are the sums of combined ranks for samples X and Y. Finally, we define the test statistic as $U = \min\{U_X, U_Y\}$. We reject H_0 (which we defined as the identical distribution of X and Y) if $U \leq U_{\text{tab}}$, where U_{tab} defines the rejection region (see e.g. Lindley and Scott [1995] for the tabulation; standard statistical software packages also provide the rejection regions).

Remark For the sample sizes n_X and n_Y greater than 10, the Mann–Whitney test can be approximated reasonably well by a normal distribution with mean $\mu_U = \frac{n_x n_y}{2}$ and variance $\sigma_U^2 = \frac{n_x n_y (n_x + n_y + 1)}{12}$. The test statistic becomes $|Z| = \frac{|U_X + U_Y|}{\sigma_U}$, and we reject H_0 of $|Z_{\text{calculated}}| \geq |Z_{\text{tab}}|$. This rule of thumb is generally not used, as most modern software packages produce exact results given the sample sizes.

Kruskal–Wallis Test

This test works for checking whether more than two samples were drawn from the same population or not. Let us consider K to be the number of comparison variables or groups, where $K > 2$. The Kruskal–Wallis test compares the medians among the K groups where we define the null hypothesis $H_0 : \mu_1 = \cdots = \mu_K$, that is, the K population medians are equal. The Kruskal–Wallis test statistic is

$$H = \left[\frac{12}{N(N + 1)} \sum_{k=1}^{K} \frac{R_k^2}{n_k} \right] - 3(N + 1) \qquad (2.83)$$

where N is the total number of samples; and n_k and R_k are the sample size and the sum of ranks for the kth group or variable, respectively.

Remark The Kruskal–Wallis test can be approximated for comparing $K \geq 3$ groups as a chi-square distribution with $K - 1$ degrees of freedom. Hence, for comparing groups in this situation we can also use the chi-square test statistic.

Wilcoxon Signed-Rank Test for Paired Data

Suppose we have paired data $(x_1, y_1), \ldots, (x_n, y_n)$, with the differences between paired observations $D = (d_1, \ldots, d_n)$, where $d_i = x_i - y_i$, $i = 1, \ldots, n$. The Wilcoxon signed-rank test considers the null hypothesis H_0 as the median difference of the paired observations being equal to zero. The goal is to analyze the signs of the differences and calculate the corresponding ranks. The sign test is also used for testing paired data, where it only takes account of the signs of the differences and not the corresponding ranks.

Spearman's Rank Correlation

Let X and Y be two random variables with an equal number of observations $n = n_x = n_y$, or equivalently, let $(x_1, y_1), \ldots, (x_n, y_n)$ be paired data. The Spearman's rank correlation coefficient is used to test the correlation between X and Y based on their corresponding ranks. The test statistic is written as

$$R_s = 1 - \frac{6 \sum_{i=1}^{n} \delta_i^2}{n^3 - n} \tag{2.84}$$

where δ_i is the difference between the ranks of the ith variables of X and Y. The value of R_s lies between -1 and $+1$, where close to -1 refers to a highly negative and $+1$ refers to a highly positive relation between X and Y.

2.3.5 Regression Model

A regression model is a way to establish a relationship between two or more variables, where a random variable Y is possibly dependent on one or more independent random variables X. It is primarily a method for establishing dependence, typically based on parametric assumptions. As a specific and useful starting point, one can consider a normal distribution to identify the fluctuations from its mean by a random amount, say ϵ. Also, to fix ideas let us assume that X denotes only one independent variable. Neither of these two assumptions – of normality and of – X being only one variable – is necessary for regression methodology, and we will relax both of these assumptions after we have gone through the basic description.

The core idea of regression is that the mean of Y conditional on observing X is a function of X. Let's write it as $E(y|X) = \mu(X)$ to imply that the mean of Y is a function of X. If this does not hold, then Y does not systematically relate to X, at least in a correlational sense. The variable X is the independent variable here. It is also described as the predictor, the explanatory variable, or the regressor, whereas the variable Y is described as the predictand, the response variable, or simply the dependent variable. A regression model also consists of a variance function, that is, $Var(y|X) = \sigma^2$. We can write a simple linear model using mathematical expectation as

$$E(y|X) = \beta_0 + \beta_1 X \tag{2.85}$$

where β_0 is called the intercept and β_1 is called the slope. In Figure 2.7, we show an elementary example of a scatter plot and the fitted regression line $\hat{Y} = \hat{\beta}_0 + \hat{\beta}_1 X$ where $\hat{Y}$ is the estimated value of Y. This $\hat{Y}$, in turn, is obtained from the estimates of the parameters $\hat{\beta}_0$ and $\hat{\beta}_1$. In this illustration, clearly $\hat{Y}$ increases with respect to X indicating a positive relationship. The whole idea of the regression analysis is to find these estimated parameter values based on a given dataset comprising X and Y variables.

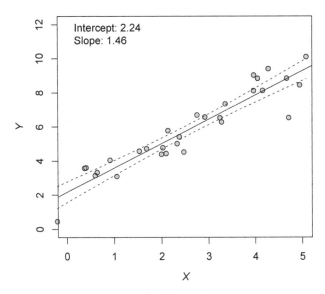

Figure 2.7 Example of scatter plot of the data points with observations on the X and Y values. We also show the best fit line (solid line) and 95% confidence band (dashed lines) obtained from the linear regression model.

Along with the above notation, let us introduce ϵ representing an error term (also known as residual or noise). In a simple form, we can write the linear model

$$Y = \beta_0 + \beta_1 X + \epsilon. \tag{2.86}$$

We can identify ϵ $(= Y - \beta_0 - \beta_1 X)$ as the deviation of the observed Y from the fitted line generated using $\beta_0 + \beta_1 X$. Note that $\beta_0 + \beta_1 X$ is the population regression line or mean function, whereas if we estimate β_0 and β_1 from sample then $\hat{\beta}_0 + \hat{\beta}_1 X$ is known as the sample regression line. Graphically, the vertical deviations from the solid straight line to the points in Figure 2.7 represent the errors. We can generalize the X to denote multiple independent variables. In matrix form, we write the more general version of the model given in Equation 2.86 in a multivariate setup (we will explain the construction below) as follows:

$$\mathbf{Y} = \mathbf{X}\beta + \epsilon. \tag{2.87}$$

Linearity in Parameters

A mean function of a regression problem can be linear or nonlinear in terms of the explanatory variables. For example, a polynomial model has a nonlinear function of X. However, the model can still be linear in its parameters. In what follows, when we say the model is linear, we imply that the regression model is *linear in parameters*.

In the following, the notation X' will denote the transpose of any vector or matrix X. Let $\mathbf{Y} = (y_1, Y_2, \ldots, Y_n)'$ be a $n \times 1$ vector. Associated with each Y_i is a $p \times 1$ vector $\mathbf{x}_i = (x_{i1}, x_{i1}, \ldots, x_{ip})'$. The following linear model describes the relationship:

$$E(y_i|X) = \mathbf{x}_i'\beta \quad i = 1, \ldots, n \tag{2.88}$$

where $\beta = (\beta_1, \ldots, \beta_p)'$ is a $p \times 1$ vector of model parameters. We assume that this linear model satisfies the assumptions of independent and identically distributed observations, and homoscedasticity of the error terms – implying that the variance is identical across observations. Let us denote it by σ_ϵ^2. Under the assumption of normality, the linear model specifies conditional distribution of the dependent variable as $Y_i \sim N(\mathbf{x}_i'\beta, \sigma_\epsilon^2)$ and we can write it as

$$Y_i = \mathbf{x}_i'\beta + \epsilon_i \quad i = 1, \ldots, n \tag{2.89}$$

where $\epsilon_1, \ldots, \epsilon_n$ are independently and identically distributed $N(0, \sigma_\epsilon^2)$ random variables. In principle, a mean function $\mu(X)$ can be nonlinear, such as a polynomial of degree p in X, and the model can be written in the following form:

$$Y_i = \beta_1 x_i + \beta_2 x_i^2 + \cdots + \beta_p x_i^p + \epsilon_i \quad i = 1, \ldots, n. \tag{2.90}$$

Our goal is to estimate the parameter vector given by β. The estimation process of the model parameters for both linear and nonlinear functional forms of regression are the same – both use ordinary least square estimation.

Ordinary Least Square Estimation for Linear Regression

The technique of least square estimation relies on a number of assumptions. For example, we require the residuals ϵ_i to have zero mean, regression variables have to be independent along with absence of outliers (nonzero finite fourth moments of X and Y), and so forth. We will review these basic necessities in more detail below. First, we set up the framework of ordinary least square estimation.

We estimate the regression parameters β by minimizing the residual sum of squares $\epsilon'\epsilon$ (Greene, 2003). From the model given by Equation 2.87, we can write $\epsilon = \mathbf{Y} - \mathbf{X}\beta$; minimizing squared is the same as minimizing the term $(\mathbf{Y} - \mathbf{X}\beta)'(\mathbf{Y} - \mathbf{X}\beta)$. Here we take the standard approach from calculus. We differentiate the sum of squares $(\mathbf{Y} - \mathbf{X}\beta)'(\mathbf{Y} - \mathbf{X}\beta)$ with respect to β and set it equal to zero to get the estimate $\hat{\beta}$.

Specifically, we can directly derive the following matrix derivative and set it equal to zero:

$$\frac{\partial \epsilon'\epsilon}{\partial \beta} = \frac{\partial}{\partial \beta}(\mathbf{Y} - \mathbf{X}\beta)'(\mathbf{Y} - \mathbf{X}\beta)$$

$$= \frac{\partial}{\partial \beta}(\mathbf{Y}'\mathbf{Y} - 2\beta'\mathbf{X}'\mathbf{Y} + \beta'\mathbf{X}'\mathbf{X}\beta)$$

$$= -2\mathbf{X}'\mathbf{Y} + 2\mathbf{X}'\mathbf{X}\beta$$
$$= 0. \tag{2.91}$$

Note that in the last equality, we are using the fact that in the optimum, the derivative has to be equal to zero. Using the last equality, we can write the estimated value of β using the hat notation as

$$\hat{\beta} = (\mathbf{X}'\mathbf{X})^{-1}\mathbf{X}'\mathbf{Y} \tag{2.92}$$

where $\mathbf{X}$ is a full rank matrix so that the term $(\mathbf{X}'\mathbf{X})$ is invertible. To check that this is indeed the minimum, we take the second derivative with respect to β and get

$$\frac{\partial^2 \epsilon'\epsilon}{\partial \beta'\partial \beta} = 2\mathbf{X}'\mathbf{X} \tag{2.93}$$

which is non-negative.

Remark   We can get the fitted values $\widehat{E(\mathbf{Y}|X)} = \mathbf{X}\hat{\beta}$ and by replacing $\hat{\beta}$ we obtain

$$\widehat{E(\mathbf{Y}|X)} = \mathbf{X}(\mathbf{X}'\mathbf{X})^{-1}\mathbf{X}'\mathbf{Y} \tag{2.94}$$

where the term $\mathbf{X}(\mathbf{X}'\mathbf{X})^{-1}\mathbf{X}'$ is commonly known as the *projection* or *hat* matrix. Under the assumption that the error term is orthogonal to X, the least square estimate of β is unbiased since

$$E(\hat{\beta}) = E((\mathbf{X}'\mathbf{X})^{-1}\mathbf{X}'\mathbf{Y})$$
$$= (\mathbf{X}'\mathbf{X})^{-1}\mathbf{X}'\mathbf{X}\beta$$
$$= \beta \tag{2.95}$$

(in the second equality, we substitute $\mathbf{Y}$ by $\mathbf{X}\beta + \epsilon$) and the corresponding variance can be written as

$$Var(\hat{\beta}) = ((\mathbf{X}'\mathbf{X})^{-1}\mathbf{X}')E(\epsilon\epsilon')((\mathbf{X}'\mathbf{X})^{-1}\mathbf{X}')'$$
$$= \sigma_\epsilon^2(\mathbf{X}'\mathbf{X})^{-1}. \tag{2.96}$$

A question can arise as to how useful this estimator is in practice in terms of its properties. It turns out that there is a very important result in the form of Gauss–Markov theorem, which states that the estimator for the regression coefficient obtained from ordinary least square estimation (OLS), that is, $\hat{\beta}$, is BLUE – an acronym that stands for the concept that it is the Best (most efficient) Linear conditionally Unbiased Estimator, under some very general conditions (see details later in this section). Although the Gauss–Markov theorem provides justification for the OLS estimate, those assumptions may not all hold in all real-life scenarios. In some cases, nonlinear and/or conditionally unbiased or other estimation methods, for example, weighted least squares (WLS), might be more efficient than OLS. We will not discuss improvements of the OLS model here. For a more detailed description of associated concepts, readers can refer to Stock and Watson (2012)

and Greene (2003). In the social sciences, it is customary to follow an econometric setup to develop the idea which relates it to identification and possible causal
interpretation of regression models, which goes beyond the statistical concept of
simply estimating the parameters. In the present context, we will not discuss causal
interpretations and will limit the discussion to parameter estimation only.

Maximum Likelihood Estimation for Linear Regression

For a regression model, under the assumption that $Y_i \sim N(\mathbf{x}'_i\beta, \sigma^2_\epsilon)$, the likelihood
function can be written as

$$L(\beta, \sigma^2_\epsilon; \mathbf{Y}, \mathbf{X}) = \prod_{i=1}^{n} (2\pi\sigma^2_\epsilon)^{-1/2} \exp\left(-\frac{1}{2\sigma^2_\epsilon}(y_i - \mathbf{x}'_i\beta)^2\right) \qquad (2.97)$$

$$= (2\pi\sigma^2_\epsilon)^{-n/2} \exp\left(-\frac{1}{2\sigma^2_\epsilon}\sum_{i=1}^{n}(y_i - \mathbf{x}'_i\beta)^2\right).$$

The log-likelihood function, in turn, would be given by

$$\log L(\beta, \sigma^2_\epsilon; \mathbf{Y}, \mathbf{X}) = -\frac{n}{2}\left(\log 2\pi + \log \sigma^2_\epsilon\right) - \frac{1}{2\sigma^2_\epsilon}\sum_{i=1}^{n}(y_i - \mathbf{x}'_i\beta)^2. \qquad (2.98)$$

Note that for the maximum likelihood estimate of β, one can simply minimize the
term $\sum_{i=1}^{n}(y_i - \mathbf{x}'_i\beta)^2$. Upon minimization, we get

$$\hat{\beta} = \frac{\sum_{i=1}^{n}(y_i - \bar{Y})(\mathbf{x}_i - \bar{X})}{\sum_{i=1}^{n}(\mathbf{x}_i - \bar{X})^2}, \qquad (2.99)$$

$$\hat{\sigma}^2_\epsilon = \frac{1}{n}\sum_{i=1}^{n}(y_i - \mathbf{x}'_i\hat{\beta})^2. \qquad (2.100)$$

We can see that the MLE for σ^2_ϵ is biased as $E(\hat{\sigma}^2_\epsilon) = \frac{n-p}{n}\sigma^2_\epsilon$. However, it is
asymptotically unbiased since when $n \to \infty$, $\frac{n-p}{n} = 1$.

Linear Regression Assumptions

Before performing a regression analysis, one has to consider a few things to make
sure that some basic assumptions are met, such as (i) linearity, (ii) independence of
the noise or error term, (iii) constant variance, that is, homoscedasticity, (iv) normality assumption of the error distribution (it is not a necessary condition though);
additionally, the data have to be free of (v) multicollinearity and (vi) outliers. While
many generalizations that can take care of these concerns have been proposed, for
the purpose of building elementary regression models, it is useful to briefly discuss
these assumptions along with some possible ways to detect and fix these issues if
they come up.

(i) **Linearity:** As we indicated earlier, the mean function or $\mu(X)$ should
be linear. A common way to identify linearity is by using graphical
plots of observed versus a predicted values, where points are expected
to be distributed symmetrically around the diagonal line. We can also
use residuals versus a predicted (or fitted) values plot. In an ideal linear
case, the points are scattered symmetrically around the horizontal line
of zero. If we observe any particular pattern or systematic noise, then
we can say that the linearity assumption no longer holds. Sometimes,
one can address this concern by rescaling the variables (either Y or X or
both).

(ii) **Independence:** If errors are not independently distributed, we cannot
perform ordinary linear regression. Time series models, which are discussed in Chapter 3, may be of use in such cases. The other approach
would be to use *generalized least squares*, which allow for correlations
across the residuals (Greene, 2003). An easy way to detect a correlation in residuals is by using autocorrelation plots. We can also use the
Durbin–Watson (DW) statistic to check for error autocorrelation at lag
1, that is, the first-order autocorrelation. We define the Durbin–Watson
statistic as follows:

$$\text{DW} = \frac{\sum_{i=2}^{n}(\epsilon_i - \epsilon_{i-1})^2}{\sum_{i=1}^{n}\epsilon_i^2} \tag{2.101}$$

where ϵ is the residual from the regression model. The common rule
of thumb is that DW values ranging from 1.5 to 2.5 indicate no strong
serial correlation for moderate sample sizes, whereas smaller values
indicate strongly positive and higher values indicate strongly negative
autocorrelations. Additionally, the errors should not be correlated with
regressors. This would introduce the problem of endogeneity and the
resulting estimates would be biased.

(iii) **Homoscedasticity:** This property refers to constant variance, that is,
σ_ϵ^2 for errors or residuals. An error series with a lack of constant variance, that is, where the variance evolves over time $-\sigma_{i\epsilon}^2, i = 1, \ldots, n$
– is called *heteroscedastic*. We can identify homoscedasticity graphi-
cally by plotting residuals versus predicted values. Similar to linearity,
for homoscedastic data we expect to see the points scattered symmetrically around the horizontal line (see the left-hand panel of Figure 2.8
for two examples). Any particular pattern of variance of the residuals as
a function of the independent variable indicates a potential violation of
the homoscedasticity assumption. Different types of transformations of
the variables or inclusion of seasonal patterns in the regression model
can alleviate the problem of heteroscedasticity.

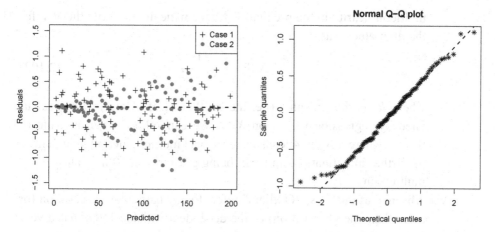

Figure 2.8 Example: panel (a) shows residual versus predicted plot for two different cases (addition and filled circle signs represent two different sets of points); panel (b) shows normal quantile–quantile plot of the residuals.

(iv) **Normality:** Sometimes the errors may not be normally distributed. While normality is not a necessary condition by any means for the regression model to work, this may create problems in estimations of the model's coefficients (i.e. β), including their statistical significance under parametric setups (e.g. MLE). A normal quantile–quantile (QQ) plot of the residuals can provide guidance on identifying the distributional pattern of the errors. A QQ plot of error points following a straight diagonal line indicates a close approximation of a normal distribution. In some cases, we can observe deviations in the bottom-left or top-right corner of the QQ plot, indicating the potential existence of outliers in the data. We can implement statistical tests to identify the normality of the error: for example, non-parametric tests such as the *Kolmogorov–Smirnov* test. The most common way to adjust a normality assumption is by scale transformation of the regression variables. Furthermore, removing outliers from the data can also help to adjust the data in favor of a normality assumption.

(v) **Absence of multicollinearity:** The existence of linear correlations between the predictors or covariates is known as multicollinearity. We assume for a linear regression that the predictors are independent. If they are not, we cannot estimate the regression coefficient β. Due to collinearity, $\mathbf{X'X}$ would not have full rank. Hence, $(\mathbf{X'X})^{-1}$ may not exist. An implication of this is that even for moderate collinearity, the variance of the estimated parameters will inflate. The *variance inflation factor* (VIF) is a statistic that can identify any serious presence of multicollinearity in the predictors. Mathematically we write the VIF for the jth predictor as

$$\text{VIF}_j = \frac{1}{1 - R_j^2} \tag{2.102}$$

where R_j^2 is the R^2 value (i.e. the coefficient of determination) calculated by regressing the X_j on $X_{j \neq k}$. We can follow a general rule of thumb, where $\text{VIF}_j > 4$ leads us to suspect multicollinearity which merits further investigation, and VIF being greater than 10 indicates strong multicollinearity.

(vi) **Absence of outliers:** "Outliers" refers to any particular observation (or more than one observation) of the dependent variable Y that has a very large deviation from the mean. Having outliers in the data often creates problem for correctly estimating the coefficients and also violates the assumption of normality. A QQ plot can provide a visual inspection to identify any potential outliers in the data (see the right-hand panel in Figure 2.8 for an example). Furthermore, one can use the *Cook's distance* statistic to diagnose the residuals for potential outliers. Mathematically, we write the Cook's distance statistic as

$$
D_j = \frac{\sum_{i=1}^{n}(\hat{Y}_i - \hat{Y}_{i(j)})^2}{p \times \frac{1}{n}\sum_{i=1}^{n}(y_i - \mathbf{x}_i'\hat{\beta})^2}
$$
$$
= \frac{\sum_{i=1}^{n}(\hat{Y}_i - \hat{Y}_{i(j)})^2}{p \times MSE} \tag{2.103}
$$

where $\hat{Y}_i$ is the ith fitted value and $\hat{Y}_{i(j)}$ is the ith fitted value after omitting the jth observation; the mean squared error is defined as $MSE = \frac{1}{n}\sum_{i=1}^{n}(y_i - \mathbf{x}_i'\hat{\beta})^2$ and $\hat{Y}_i = \mathbf{x}_i'\hat{\beta}$. Cook's distance having a magnitude of 1 or more indicates a potential existence of outliers in the data.

Often we observe a similarly large deviation for a particular point (or points) of the predictor, that is, X, from its mean, which also influences the regression model estimation. This is known as *leverage* in regression analysis. Scatter plots of X and Y variables allow us to visualize the presence of leverage. Let x_{ij} be the predictor for the jth variable where $j = 1, \ldots, p$, and the ith observation where $i = 1, \ldots, n$. A rule of thumb for detecting leverage points is that x_{ij} is treated as a leverage point if $x_{ij} > (3 \times \bar{x}_j)$, where $\bar{x}_j = \sum_{i=1}^{n} x_{ij}/n$.

Hypothesis Tests for Linear Regression Parameters

Here, we represent statistical hypothesis tests and confidence intervals for the regression coefficient $\beta = (\beta_1, \ldots, \beta_p)'$. Testing a single coefficient, that is,

β_j, $j = 1, \ldots, p$, we can provide a decision for corresponding predictor $\mathbf{x}_j = (x_{1j}, \ldots, x_{nj})'$ for n observations. For a two-sided test we write the null hypothesis $H_0 : \beta_j = \beta_{0j}$ and the corresponding alternative hypothesis $H_1 : \beta_j \neq \beta_{0j}$. To perform the test we estimate the coefficients, that is, $\hat{\beta}_j$, and the corresponding standard error, that is, $se(\hat{\beta}_j)$. Hence, we compute the t-statistic

$$t = \frac{\hat{\beta}_j - \beta_{0j}}{se(\hat{\beta}_j)}. \tag{2.104}$$

We would reject the null hypothesis if $|t_{cal}| > t_{tab}$, where t_{cal} is the calculated t-value and t_{tab} is the table value at 5% significance level. In most cases we test $H_0 : \beta_j = 0$ with the alternative $H_0 : \beta_j \neq 0$. For testing a single coefficient we write the 95% two-sided confidence interval as

$$\left(\hat{\beta}_j - t_{tab} \times se(\hat{\beta}_j), \hat{\beta}_j + t_{tab} \times se(\hat{\beta}_j) \right). \tag{2.105}$$

For testing all coefficients together we use the F-statistic, which can show the overall significance of the regression model. Suppose we want to test the null hypothesis

$$H_0 : \beta_1 = 0, \beta_2 = 0, \ldots, \beta_p = 0$$

with the alternative H_1 : at least one β_j is zero. Hence, we write the F-test statistic for p and $n - p$ degrees of freedom as

$$F = \frac{SSR/p}{SSE/(n - p)} \tag{2.106}$$

where $SSR = \sum_{i=1}^{n}(\hat{Y}_i - \bar{Y})^2$ is the regression sum of squares and $SSE = \sum_{i=1}^{n}(y_i - \hat{Y}_i)^2$ is the error sum of squares. We can get the total sum of squares TSS by combining SSR and SSE, that is, $TSS = SSR + SSE$, and mathematically we write $TSS = \sum_{i=1}^{n}(y_i - \bar{Y})^2$. The terms TSS, SSR, and SSE directly provide the construction of ANOVA (analysis of variance) for the linear regression model.

Linear Regression: Economic Complexity and Prosperity

Here we take a simple case of multivariate regressions to describe an application of the technique. Hidalgo and Hausmann (2009) constructed a measure of economic complexity based on the diversity of knowledge capital that is used for producing tradable goods. It is intuitive that possibly richer countries would be closer to the knowledge frontier and, hence, the economic complexity index would be positively correlated with the per capita GDP of countries. One way to capture this idea is to estimate a univariate regression model according to Equation 2.86 on a set of countries. We collected per capita GDP data on 50 economies from the World Bank and their complexity index in 2017 from the Atlas of Economic Complexity. The resulting scatter plot and the linear regression line are shown in Figure 2.9.

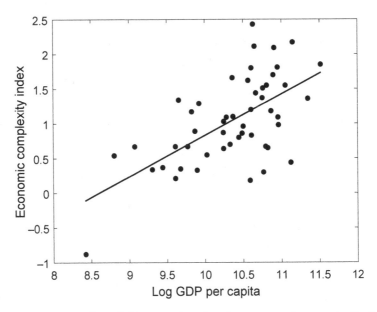

Figure 2.9 Scatter plot of 50 countries showing economic complexity index and per capita GDP. The linear fit shows a strong relationship between the two variables. The complexity measure is computed and produced following the methodology proposed by Hidalgo and Hausmann (2009).

The OLS estimate shows that the estimated coefficient β_1 is equal to 0.59 with standard error equal to 0.11. The t-statistic is 5.43 and the coefficient is significant at 1% (in fact the p-value is on the order of 10^{-6}). The model has an R-square of 0.38. Next, we augment the model with a measure of trade openness (exports plus imports divided by GDP) for each country. we do this on the basis that more engagement in global trade would possibly positively impact technology development in a given country. With this addition, the resulting model has two variables on the right-hand side. The OLS estimate of the augmented model shows that the GDP per capita variable remains significant with a slight modification in the coefficient value (0.61) and standard error (0.12). However, the newly added variable trade openness has a coefficient of -0.06 with a standard error of 0.14 and p-value of 0.64. Therefore, the new variable does not have any significant relationship with economic complexity. The R-square is 0.38 and the adjusted R-square is 0.35, indicating that adding the new variable does not have any additional explanatory power, and in fact, when the R-square is penalized for the number of explanatory variables, it actually goes down. A final point is to be noted here. In the present context, we cannot resolve the problem of causality. High GDP per capita may lead to development of more complex technology and more complex technology, in turn, may lead to high GDP per capita. Modern econometrics has suitable toolkits for analyzing casual channels which fall outside the scope of the present discussion. Interested readers may consult Angrist and Pischke (2008) (see also Imbens and Rubin [2015]).

2.3.6 Non-normal Data Model

Here we consider models where the response variables may not be normally distributed. A common example of such models is the generalized linear model (GLM), which allows the response variable to follow a larger class of distributions called the exponential family of distributions (Sundberg, 2019).

Exponential Family of Distributions

The exponential family of distributions contains a large number of commonly known distributions including normal, exponential, Poisson, gamma, Bernoulli/binomial, etc. The general form of the probability density function (pdf) is

$$f_Y(y; \theta, \xi) = \exp\left(\frac{\theta y - b(\theta)}{a(\xi)} + c(y, \xi)\right) \tag{2.107}$$

where θ represents the *natural* or *canonical* parameters, $a(.)$, $b(.)$, and $c(.,.)$ are known functions, and ξ denotes *nuisance* parameters which are usually assumed to be known. It can be shown with some algebra that the mean $E(y)$ would be equal to $b'(\theta)$ and variance $Var(y)$ would be equal to $a(\xi)b''(\theta)$, where $b'(\theta)$ and $b''(\theta)$ are the two consecutive derivatives of $b(\theta)$.

For example, for a Bernoulli distribution with parameters μ, we write the pdf as

$$f_Y(y) = \mu^y (1 - \mu)^{1-y} \quad \mu \in (0, 1)$$

$$= \exp\left(y \log \frac{\mu}{1 - \mu} + \log(1 - \mu)\right) \tag{2.108}$$

where $\theta = \log\frac{\mu}{1-\mu}$, $b(\theta) = \log(1 + e^\theta)$, $a(\xi) = 1$, and $c(y, \xi) = 0$. Hence, we can take the derivatives and derive the first two moments as

$$E(y) = \frac{e^\theta}{1 + e^\theta} = \mu,$$

$$Var(y) = \frac{e^\theta}{(1 + e^\theta)^2} = \mu(1 - \mu).$$

Generalized Linear Model

With the above description of the exponential family of distributions, we can now define the GLM setup. Before we describe the model, we should add a small note regarding the notation b'. In the following discussion of the GLM, we use the prime notation for both the transpose of a matrix and a derivative. It is used as a derivative only for the function denoted by $b(.)$. In all other cases, the prime notation is used for denoting a transpose.

Let $Y = (y_1, \ldots, Y_n)'$ be the set of independent random variables with the same distribution as the exponential family of distributions. We write the joint density function as

$$f_Y(y; \theta, \phi) = \prod_{i=1}^{n} f_{Y_i}(y_i; \theta_i, \phi_i)$$

$$= \exp \left(\sum_{i=1}^{n} \frac{y_i \theta_i - b(\theta_i)}{a(\phi_i)} + \sum_{i=1}^{n} c(y_i, \phi_i) \right). \tag{2.109}$$

Now, under the modeling structure with p predictors $x_i = (x_{i1}, \ldots, x_{ip})'$, we write the generalized linear model (GLM) through a linear predictor η_i as

$$\eta_i = \sum_{k=1}^{p} \beta_k x_{ik} = x_i'\beta$$

where i varies from 1 to n. In matrix notation, we write it as

$$\eta = X\beta \tag{2.110}$$

where X is called the $n \times p$ design matrix, and $\eta = (\eta_1, \ldots, \eta_n)'$. The link between the response Y and the linear predictor η is established using a *link function* $g(\mu)$ where $\mu \equiv E(Y)$. Therefore, we can write

$$g(E(Y)) = g(\mu) = \eta = X\beta.$$

Note that for an exponential family of distribution $E(Y) = b'(\theta)$. So we get the following:

$$\mu = E(Y) = b'(\theta).$$

This equation directly implies

$$\theta = b'^{-1}(\mu)$$
$$= b'^{-1}\left(g^{-1}(X\beta)\right). \tag{2.111}$$

Here, the function g and b'^{-1} are identical, that is, $g(\mu) \equiv b'^{-1}(\mu)$, which is known as the *canonical* link function. Therefore, the canonical parameter is equal to the component capturing the linear predictor. In short, one can write $\theta = X\beta$.

Remark   Note that for an identity link, the GLM becomes a linear regression model. For other canonical links, the model also becomes a linear model through the linear predictor, that is, $\eta = X\beta$. Hence, we can implement maximum likelihood estimation to obtain estimates of the parameter β. However, the score function

Table 2.3 *Canonical link functions for generalized linear models*

	$b(\theta)$	$b'(\theta) \equiv \mu$	$b'^{-1}(\mu) \equiv \theta$	Link	Name
Normal	$\frac{\theta^2}{2}$	θ	μ	$g(\mu) = \mu$	Identity link
Poisson	e^{θ}	e^{θ}	$\log \mu$	$g(\mu) = \log \mu$	Log link
Binomial	$\log(1 + e^{\theta})$	$\frac{e^{\theta}}{1+e^{\theta}}$	$\log \frac{\mu}{1-\mu}$	$g(\mu) = \log \frac{\mu}{1-\mu}$	Logistic link

using the linear predictor becomes a system of nonlinear equations (except for the identity link, that is, for normal distribution) and we do not have any analytic solution for this. Hence, we use numerical methods such as the Newton–Raphson method to estimate β. We can also use an iterative minimization of the weighted sum of squares to get the estimate of β, which is known as the iterative reweighted least square estimate. For details on the GLM estimation, see Dobson and Barnett (2018) and McCullagh (2018).

Remark The GLM can be extended as an additive model where, similarly to semi-parameteric regression, additive terms are used for the predictors. We denote it as the generalized additive model (GAM), where the errors are non-normal, and we use the link function if the predictand follows in the exponential family of distribution. Therefore, we can write the linear predictor for the GAM:

$$\eta_i = f(x_i) = \sum_{k=1}^{K} \gamma_k b_k(x_i). \tag{2.112}$$

On a related note, in Section 4.3.2 we will analyze a specific form of this class of models – logistic regression, in the context of classification. For further details on generalized linear models, one can consult Wood (2017), Hastie and Tibshirani (1990), and Hastie et al. (2009).

2.3.7 Model Selection

The concept of model selection deals with the aspects of the performance of models given data where explanatory variable selection often plays a key role. Sometimes a regression model may contain a number of possible explanatory or independent variables, which may introduce challenges to fulfilling the assumptions of a regression model. The main concept of model selection lies in choosing a subset of predictors that can provide relatively better performance of the model. There are some criteria available to evaluate models in terms of their performance (Casella and Berger, 2021).

Deviance

Deviance is defined to be a function of the likelihood function, where usually we take a logarithmic form of the likelihood and hence write

$$\text{deviance} = 2 \times \log L(\theta_s) - 2 \times \log L(\theta) \qquad (2.113)$$

where the function $L(\theta_s)$ represents the likelihood of the saturated model which has parameters for all combinations of the covariates. The term $L(\theta)$ defines the likelihood for a given model with parameters θ.

Coefficient of Determination R^2

This coefficient represents the amount of variation explained by the model's inputs and is defined as

$$R^2 = \frac{\sum_i (\hat{Y}_i - \bar{Y})^2}{\sum_i (y_i - \bar{Y})^2} = 1 - \frac{\sum_i (y_i - \hat{Y})^2}{\sum_i (y_i - \bar{Y})^2} \qquad (2.114)$$

where $\sum_i (y_i - \bar{Y})^2$ is denoted as the sum of squares due to the total variation (SST), $\sum_i (y_i - \hat{Y}_i)^2$ represents the sum of squares due to the residuals or errors (SSE), and $\sum_i (\hat{Y}_i - \bar{Y})^2$ represents the sum of squares due to regression (SSR).

Remark Even if we add predictor or explanatory variables that are irrelevant to the regression, they would still (weakly) increase the value of R^2. To counteract such increases, we use an adjustment of the R^2 by penalizing for the number of explanatory variables:

$$R^2_{adj} = 1 - \frac{SSE/(n - p - 1)}{SST/(n - 1)} \qquad (2.115)$$

where n is the number of observations and p is the number of predictors. R^2 is between 0 and 1. A value close to 1 yields a regression model that explains a large amount of variation in the response variable around the mean. A value close to 0 indicates the opposite – a low amount of variation being explained. On the other hand, theoretically it is possible that R^2_{adj} may turn out to be negative – when the sample size is very small relative to the number of predictors. Such negative values can be ignored and are indicative of too many explanatory variables. Accordingly, a more economical model should be considered.

The coefficient of determination is an easily interpretable summary of the goodness of fit of a regression model. However, it does not quantify the complexity of the model, and an increase in the R^2 does not necessarily indicate that the model is better, as adding more variables to the model is bound to weakly increase R^2 even if it does not truly explain the variation in the endogenous variable.

We will revisit this idea in Chapter 4 to discuss the explanatory power of a model in the context of machine learning.

Mallows' C_p

Mallows' C_p combines deviance and the number of regression parameters p and expresses a measure of goodness of fit and complexity in the model selection criterion:

$$C_p = D + 2p\sigma^2 \tag{2.116}$$

Here, D refers to the deviance defined above and $2p\sigma^2$ captures the complexity of the model. Here, "complexity" refers to how many parameters there are in the model and what forms they take; the term here does not refer to features of complex systems.

In most cases, σ^2, that is, variance, is replaced by the estimated variance obtained from the data. The smaller the value of C_p, the better the model. Although it is relatively easy to compare regression models using their corresponding C_ps, it is essentially an ad hoc summary measure.

Akaike Information Criterion (AIC)

The Akaike information criterion is based on goodness of fit and a corresponding penalty for model complexity. Suppose we have $\theta = (\theta_1, \ldots, \theta_p)'$, p model parameters, and a corresponding likelihood function $L(\theta)$; hence, we write the AIC as

$$\text{AIC} = 2[-\log L(\theta) + p]. \tag{2.117}$$

For a linear regression model we can write the AIC as

$$\text{AIC} = n \log \left(\frac{\text{SSE}}{n} \right) + 2p. \tag{2.118}$$

Here, the term $n \log \left(\frac{\text{SSE}}{n} \right)$ is the goodness of fit and $2p$ captures the model's complexity. Note that the smaller the AIC is, the better the regression model. The well-known leave-one-out cross-validation approach is asymptotically equivalent to AIC in a statistical model (Watanabe and Opper, 2010).

Remark For a small sample size, the AIC needs a correction for its over-fitting nature due to the small amount of model complexity. Hence, we write the corrected AIC:

$$\text{AIC}_c = AIC + \frac{2p(p+1)}{n-p+1}. \tag{2.119}$$

Usually, we use AIC_c if $\frac{n}{p} > 40$.

Bayesian Information Criterion (BIC)

The Bayesian information criterion is similar to the AIC, except that the model complexity term is defined $p \log n$. Thus we write

$$\text{BIC} = -2 \log L(\theta) + p \log n. \tag{2.120}$$

BIC favors more parsimonious models than the AIC because it imposes a heavier penalty for larger models.

2.4 Bayesian Statistics and Inference

In this section, we will discuss the conceptual framework of Bayesian statistics by emphasizing their applicability and attractiveness for solving data problems. In Bayesian inference we combine prior information with sample data to gain *a posteriori* information on the subject matter. The basic idea is to estimate a distributional pattern of the model's parameters by considering prior and likelihood information coming from the data. The Bayesian approach gives us information on the basis of beliefs about the classical inference relies on the frequency of events (Gelman et al., 2013). In comparison with the classical approach, generally Bayesian approaches have better small-sample properties. However, the computational cost is often substantially higher than in the classical approach. Technical differences aside, more broadly, the Bayesian paradigm is even philosophically different from the frequentist paradigm. Below we will directly delve into the technical description. Interested readers can refer to Gelman and Shalizi (2013) for a discussion on the philosophical aspect of this paradigm.

2.4.1 Bayes' Theorem and Prior Elucidation

We have already discussed Bayes' theorem (Equation 2.4) in Section 2.2.1. Following the same idea, we can write the rule of updating beliefs about parameters θ based on observations y as the following:

$$\pi(\theta|y) \propto \pi(\theta)p(y|\theta) \tag{2.121}$$

where $\pi(\theta)$ refers to the prior distribution of parameter θ, and $p(y|\theta)$ is the likelihood function. This way of articulating Bayes' theorem refers to an unnormalized posterior density where the proportionality operation has been used. We can write the posterior density in the expanded form as follows:

$$\pi(\theta|y_1,\ldots,y_n) = \frac{p(y_1,\ldots,y_n|\theta)\pi(\theta)}{p(y)} = \frac{p(y_1,\ldots,y_n|\theta)\pi(\theta)}{\int_\theta \pi(\theta)p(y_1,\ldots,y_n|\theta)d\theta}.$$

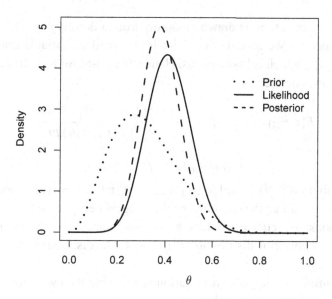

Figure 2.10 Bayesian concept: Prior, likelihood, and posterior distributions for some parameter θ.

The integral in the denominator does not depend on θ, and thus it can be treated as a constant. We can ignore it for getting posterior inference. Figure 2.10 shows an example of simulated density plots for likelihood, prior, and posterior distributions, where we see that the posterior distribution is shifted a bit (in this case to the left) compared to the density of data likelihood with the influence from the prior. We also see that the posterior density has less spread compared to the prior and likelihood densities as it incorporates information from both prior and likelihood. This kind of updating is often described as Bayes' *rule*.

Remark Following Bayes' rule, we construct a simple model of Bayesian learning. Suppose we evaluate θ based on only one observation y_1:

$$\pi(\theta|y_1) \propto p(y_1|\theta)\pi(\theta).$$

Next, suppose we have a second observation y_2. We update the probability in the following way:

$$\pi(\theta|y_1, y_2) \propto p(y_2|\theta)p(y_1|\theta)\pi(\theta)$$
$$\propto p(y_2|\theta)\pi(\theta|y_1). \tag{2.122}$$

In this way we can successively iterate and gain new estimates of the model's parameters dynamically with the availability of newer and newer data.

Let $y_1, \ldots, y_n$ be a sample drawn randomly from a density $f(y, \theta)$, where θ represents parameters. We assume that θ itself is a random variable drawn from Θ with density $p(\theta)$ which is known to us. Hence, the posterior Bayesian estimator of a function $\tau(\theta)$ is defined as

$$E[\tau(\theta)|y_1, \ldots, y_n] = \frac{\int \tau(\theta) \left[\prod_i f(y_i, \theta) \right] \pi(\theta) d\theta}{\int \left[\prod_i f(y_i, \theta) \right] \pi(\theta) d\theta}.$$

A Note on Prior Distribution

A prior distribution $\pi(\theta)$ is said to be *proper* if $\int_\theta \pi(\theta) d\theta = 1$. An *improper* prior distribution may violate the condition of the integral being equal to 1: for example, it may be a constant, $\pi(\theta) = k$, where $k > 0$. Sometimes we use an improper prior in updating the posterior distribution if the posterior distribution is proper where $\int_\theta \pi(\theta|y) d\theta < \infty$.

Prior ignorance: If no prior information is available then we define the prior as prior ignorance or non-informative prior.

Jeffrey's rule for non-informative prior: Let $\pi(\theta)$ be the prior distribution for parameter θ. Jeffrey's rule suggests that the non-informative prior is proportional to the square root of Fisher's information

$$\pi(\theta) \propto [I(\theta)]^{1/2} \quad \text{for} \quad \theta \in \Theta$$

where $I(\theta) = -E\left[\frac{\partial^2}{\partial \theta^2} \log p(y|\theta) \right]$ is the Fisher's information.

Example 2.13 Let $Y_1, \ldots, Y_n$ be a random sample from a Poisson distribution with parameter θ. We write the joint density as

$$p(y|\theta) = \frac{1}{\prod_{i=1}^n Y_i} e^{(-n\theta)} \theta^{\left(\sum_{i=1}^n Y_i \right)}. \tag{2.123}$$

Now we calculate Fisher's information $I(\theta)$ to identify Jeffrey's non-informative prior. We take the logarithm of the joint density to get

$$\log p(y|\theta) = -n\theta + \sum_{i=1}^n y_i \log \theta - \log \left(\prod_{i=1}^n Y_i \right). \tag{2.124}$$

Taking two consecutive derivatives, we get

$$\frac{\partial}{\partial \theta} \log p(y|\theta) = -n + \frac{\sum_{i=1}^n Y_i}{\theta}$$

$$\frac{\partial^2}{\partial \theta^2} \log p(y|\theta) = -\frac{\sum_{i=1}^n Y_i}{\theta^2},$$

Thus, we write the Fisher's information as

$$I(\theta) = E\left(-\frac{\partial^2}{\partial\theta^2}\log p(y|\theta)\right) = E\left(\frac{\sum_{i=1}^{n} Y_i}{\theta^2}\right) = \frac{n\theta}{\theta^2} = \frac{n}{\theta}. \quad (2.125)$$

Hence, Jeffrey's non-informative prior is

$$\pi(\theta) \propto [I(\theta)]^{1/2}$$
$$\propto \theta^{-1/2}. \quad (2.126)$$

Using the Jeffrey's non-informative prior, the posterior can be written as

$$p(\theta|y) \propto p(y|\theta)\pi(\theta) = e^{(-n\theta)}\theta^{(\sum_{i=1}^{n} Y_i)} \times \theta^{-1/2}$$
$$\propto e^{(-n\theta)}\theta^{(\sum_{i=1}^{n} Y_i - 1/2)}. \quad (2.127)$$

The posterior is a gamma distribution with parameters $\sum_{i=1}^{n} Y_i - 1/2$ and n.

Conjugate prior family: If $\pi(\theta)$ is a member of some parametric family F and the posterior distribution of θ given by $\pi(\theta|y)$ is also a member of the same parametric family F, then $\pi(\theta)$ is a conjugate prior for $\pi(y|\theta)$.

Example 2.14 Let $Y_1, \ldots, Y_n$ be a sample randomly drawn from a Bernoulli distribution with parameter $\theta \in (0, 1)$. The posterior density of θ given observation y is

$$\pi(\theta|y) \propto p(y|\theta)\pi(\theta)$$
$$\propto \theta^{\sum_{i=1}^{n} Y_i}(1 - \theta)^{n-\sum_{i=1}^{n} Y_i}\pi(\theta). \quad (2.128)$$

Here we obtain the conjugate prior by identifying the kernel of the likelihood function. We write the density function as a function of the beta distribution family with parameters α and β, that is,

$$\pi(\theta) = \theta^{\alpha-1}(1 - \theta)^{\beta-1}.$$

This allows us to write the posterior distribution as

$$\pi(\theta|y) \propto \theta^{\sum_{i=1}^{n} Y_i}(1 - \theta)^{n-\sum_{i=1}^{n} Y_i} \times \theta^{\alpha-1}(1 - \theta)^{\beta-1}$$
$$\propto \theta^{\sum_{i=1}^{n} Y_i+\alpha-1}(1 - \theta)^{n-\sum_{i=1}^{n} Y_i+\beta-1}. \quad (2.129)$$

Therefore, $\pi(\theta|y)$ also follows beta distribution with parameters $\sum_{i=1}^{n} Y_i + \alpha$ and $n - \sum_{i=1}^{n} Y_i + \beta$, which belongs to the same family of distribution used for the prior. Hence, we can say that the beta family is conjugate for the Bernoulli model.

2.4.2 Bayesian Inference

Bayesian Hypothesis Testing

Let us start with a null hypothesis $H_0 : \theta = \theta_0$ and corresponding alternative $H_1 : \theta \neq \theta_0$ where θ is the parameter of interest. Under the Bayesian paradigm, hypothesis testing involves utilization of a prior for the H_0, that is, $\pi(H_0)$, and for θ, that is, $\pi(\theta)$, to compute the posterior $\pi(H_0|y)$. Using Bayes' theorem, we write

$$\pi(H_0|y) = \frac{p(y|H_0)\pi(H_0)}{p(y|H_0)\pi(H_0) + p(y|H_1)\pi(H_1)} \tag{2.130}$$

A common consideration of priors for H_0 and H_1 is $\pi(H_0) = \pi(H_1) = 1/2$. Hence, we can write the posterior in the following form:

$$\pi(H_0|y) \quad = \frac{p(y|H_0)}{p(y|H_0) + p(y|H_1)} \tag{2.131}$$

$$= \frac{p(y|H_0)}{p(y|H_0) + \int \pi(\theta)p(y|\theta)d\theta}. \tag{2.132}$$

Here $\pi(\theta)$ should be a proper prior specification. Otherwise, the expression $\int \pi(\theta)p(y|\theta)d\theta$ may not be well defined.

Credible Regions

Under the Bayesian framework, we can define credible regions for parameter θ. Let us denote C as the credible region such that $P(\theta \in C|y) = 1 - \alpha$ where α is the level of significance. Upon such a specification, C denotes the $100(1 - \alpha)\%$ credible region for θ. This C is termed the *highest posterior density credible region* if $\pi(\theta|y) \geq \pi(\tau|y)$ where $\theta \in C$ and $\tau \notin C$.

Posterior Predictive Distribution

Let $Y_1, \ldots, Y_n$ be a sample randomly drawn from a distribution $f(y, \theta)$ with parameter θ. To make predictions, the goal is to obtain the distribution of Y_{n+1} given $Y_1, \ldots, Y_n$; that is, we already know about the samples $Y_1, \ldots, Y_n$. The posterior predictive distribution is defined as

$$p(y_{n+1}|y_1, \ldots, y_n) = \int p(y_{n+1}|\theta)\pi(\theta|y_1, \ldots, y_n)d\theta, \tag{2.133}$$

where $\pi(\theta|y_1, \ldots, y_n)$ denotes the posterior distribution obtained from observed realizations of $Y_1, \ldots, Y_n$.

2.4.3 Bayesian Computation

Monte Carlo Simulation

As we have already seen above, Bayesian inference involves integrals and often the integrals are not easy to solve. The Monte Carlo method is one of the most fundamental numerical computation methods that eases the computational burden. The basic concept of the method lies in repeated sampling from a density to learn about the variable or parameter of interest. In this approach, we generate a sample of the parameter θ values ($\{\theta^{(1)}, \ldots, \theta^{(m)}\}$) from the posterior distribution and use sample statistics to approximate the $\pi(\theta|y)$. With this method, we can approximate the following integral with a function $g(\theta)$:

$$\int g(\theta)\pi(\theta|y)d\theta = E[g(\theta)|y]. \tag{2.134}$$

The starting point is to estimate the integral by simulating $\{\theta^{(1)}, \ldots, \theta^{(m)}\}$ with the sample average

$$\hat{E}[g(\theta)|y] = \frac{1}{m}\sum_{j=1}^{m} g\left(\theta^{(j)}\right). \tag{2.135}$$

The sample average $\hat{E}[g(\theta)|y]$ converges to $E[g(\theta)|y]$ following the law of large numbers. Therefore, we can write it as the population analogue as

$$E[g(\theta)|y] \approx \frac{1}{m}\sum_{j=1}^{m} g\left(\theta^{(j)}\right). \tag{2.136}$$

Naturally, the estimate becomes more accurate for larger values of m as the approximation gets better asymptotically.

Markov Chain Simulation

The Monte Carlo method has a drawback in terms of getting independent samples directly from $\pi(\theta|y)$. One can overcome this problem by utilizing a Markov chain in conjunction with the Monte Carlo method. This leads to a Markov chain simulation, also known as the Markov Chain Monte Carlo (MCMC) method. This method involves sampling $\theta^{(j+1)}$ from a transition kernel $\kappa\left(\theta|\theta^{(j)}\right)$, where $j = 1, 2, \ldots, m$. Such sampling implies that $\theta^{(j+1)}$ depends only on $\theta^{(j)}$ and not on $\theta^{(j-1)}$, that is,

$$\kappa\left(\theta^{(j+1)}|\theta^{(j)}, \theta^{(j-1)}, \theta^{(j-2)}, \ldots\right) = \kappa\left(\theta^{(j+1)}|\theta^{(j)}\right). \tag{2.137}$$

The convergence of the process is based on finding a stationary distribution. It can be shown that under some fairly general conditions, an irreducible Markov chain will have a stationary distribution which would also be unique.

Metropolis–Hastings Algorithm

A general and flexible Markov Chain Monte Carlo method comes in the form of the Metropolis–Hastings algorithm. Here the idea is to sample a proposal distribution or jumping distribution $J(\theta^*|\theta^{(j-1)})$ at iteration j. The steps of the Metropolis–Hastings algorithm as follows.

1. Initiate a starting point for the parameter at θ^0, at iteration $j = 1$.
2. Sample θ^* from the proposal distribution $J(\theta^*|\theta^{(j-1)})$ at iteration j.
3. Generate a number to be used as a probability of updating:

$$\alpha(\theta^{(j-1)}, \theta^*) = \min\left\{1, \frac{\pi(\theta^*|y)J(\theta^*|\theta^{(j-1)})}{\pi(\theta^{(j-1)}|y)J(\theta^{(j-1)}|\theta^*)}\right\}$$

$$= \min\left\{1, \frac{\pi(\theta^*)p(y|\theta^*)J(\theta^*|\theta^{(j-1)})}{\pi(\theta^{(j-1)})p(y|\theta^{(j-1)})J(\theta^{(j-1)}|\theta^*)}\right\}.$$

4. Generate a uniformly distributed random number: $U \sim \text{uniform}(0, 1)$.
5. If $U \leq \alpha(\theta^{(j-1)}, \theta^*)$, update the parameter following $\theta^{(j)} \leftarrow \theta^*$. Otherwise, continue with the same value $\theta^{(j)} \leftarrow \theta^{(j-1)}$.

For the Metropolis–Hastings algorithm, a normalizing constant in posterior distribution $\pi(\theta|y)$ is not required because it cancels in the ratio.

Remark   The Metropolis–Hastings algorithm boils down to the Metropolis algorithm when the proposal distribution is symmetric in nature, that is, $J(\theta^*|\theta^{(j-1)}) = J(\theta^{(j-1)}|\theta^*)$. In this case, we write the probability as

$$\alpha(\theta^{(j-1)}, \theta^*) = \min\left\{1, \frac{\pi(\theta^*|y)}{\pi(\theta^{(j-1)}|y)}\right\}.$$

Other than that, we use the exact same steps described for the Metropolis–Hastings algorithm.

Gibbs Algorithm

Implementation of the Gibbs algorithm allows us to obtain a p-dimensional posterior distribution of parameters, for example, $\theta = (\theta_1, \ldots, \theta_p)'$, by iteratively sampling from each conditional posterior distribution. It is one of the best choices for Bayesian conjugate models. This algorithm updates the iterative chains by one component at a time: that is, at any particular iteration j, we write

$$\theta_1^{(j)} \sim \quad \pi(\theta_1|\theta_2^{(j-1)}, \theta_3^{(j-1)}, \ldots, \theta_p^{(j-1)}, \mathbf{y})$$

$$\theta_2^{(j)} \sim \quad \pi(\theta_2|\theta_1^{(j-1)}, \theta_3^{(j-1)}, \ldots, \theta_p^{(j-1)}, \mathbf{y})$$

$$\vdots$$

$$\theta_p^{(j)} \sim \quad \pi(\theta_p|\theta_1^{(j-1)}, \theta_2^{(j-1)}, \ldots, \theta_{p-1}^{(j-1)}, \mathbf{y}).$$

In Figure 2.11, we show the convergence of the MCMC chains for the parameter estimates for gamma and lognormal distributions fitted to the consumption data in Figure 2.6.

Similarly, in Figure 2.12, we show the convergence of a single-chain MCMC with two priors. We have assumed that both the intercept and the slope parameters

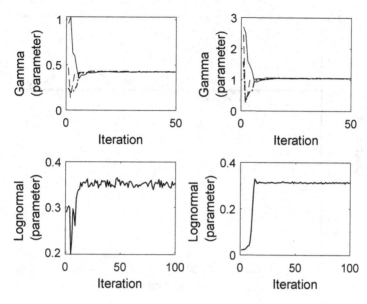

Figure 2.11 We estimate the parameters of the gamma and lognormal distributions fitted to the empirical distribution in Figure 2.6, with a Bayesian approach. The top panels show three starting points for running the MCMC chains, which very quickly converge. We use proper and non-informative (flat) conjugate prior distributions to get the posterior samples of the distributions. The bottom panels show the evolution of a single MCMC chain for the two parameters of the lognormal distribution.

would have a prior in the form of $N(\mu, \sigma)$. In this particular case, we have used two different scenarios for the hyper-prior parameters, where first we considered a large $\sigma = 10,000$ and then we ran the model again with a small $\sigma = 0.1$. The results show that use of large σ leads to a rapid convergence of the posterior MCMC samples, whereas for a small σ value the MCMC fluctuates with less chance of convergence for our model based on the data.

2.4.4 Bayesian Hierarchical Model

A Bayesian hierarchical model captures the joint distribution of *data, process,* and *parameters* (Berliner, 1996; Cressie and Wikle, 2015; Gelfand, 2012). In the first level of the hierarchy, a data model is developed considering a latent true process and model parameters. The second level of the hierarchy is the process model considering the model parameters. The third level is the parameter model itself. Intuitively, we can write the levels as follows. In the first stage, a {data model} gives us the description of data conditional on process and parameters – [data|process, parameters]. This can be updated to get to the second stage of the

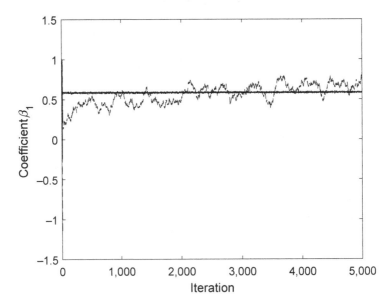

Figure 2.12 Convergence of the MCMC chains (with high and low variance of the priors for the slope coefficient β_1) for the linear model estimates of the relationship between economic complexity and per capita GDP as in Figure 2.9. As the figure shows, the estimate with large variance prior converges quickly and remains flat, whereas the estimate with small variance prior exhibits a substantial amount of fluctuation.

{process model}, which gives us [process|parameters]. Finally, the third stage in the form of a {parameter model} gives us the [*parameters*]. Under the Bayesian paradigm, we can write the corresponding hierarchical structure as follows:

$$[\text{process, parameters}|\text{data}] \propto [\text{data}|\text{process, parameters}] \times$$

$$[\text{process}|\text{parameters}] \times [\text{parameters}].$$

Let us explain the hierarchy using the regression model defined in Section 2.3.5. Consider a linear model $\mathbf{Y} \sim N(\mu, \sigma_\epsilon^2 \mathbf{I})$ where μ is the true process which follows a normal distribution, that is, $\mu \sim N(\mathbf{X}\beta, \Sigma_w)$. Here the term σ_ϵ^2 is the variance of the white noise and the term $\Sigma_w = \sigma_w^2 \mathbf{S}_w$ is the variance–covariance term associated with the true process μ. We can frame this model using the Bayesian hierarchy as follows:

$$\pi(\mu, \beta, \sigma_\epsilon^2, \sigma_w^2 | \mathbf{Y}, \mathbf{X}) \propto \pi(\mathbf{Y}|\mathbf{X}, \mu, \beta, \sigma_\epsilon^2, \Sigma_w) \times$$

$$\pi(\mu | \mathbf{X}, \beta, \Sigma_w) \times \pi(\beta, \sigma_\epsilon^2, \Sigma_w).$$

In the regression context, the distribution of $\mathbf{X}$ is not informative in terms of model parameters, and hence we can write the joint likelihood as $\pi(\mathbf{Y}, \mathbf{X}|\mu, \beta, \sigma_\epsilon^2, \sigma_w^2)$ $= \pi(\mathbf{Y}|\mathbf{X}, \mu, \beta, \sigma_\epsilon^2, \Sigma_w)$: see, for example, Gelman et al. (2013).

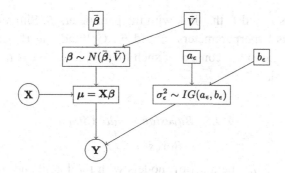

Figure 2.13 Directed acyclic graph of the Bayesian regression model.

Now suppose that the prior distributions are independent. Then we can write the joint priors as $\pi(\beta, \sigma_\epsilon^2, \Sigma_w) = \pi(\beta) \times \pi(\sigma_\epsilon^2) \times \pi(\Sigma_w)$. Considering conjugacy, we write $\pi(\beta) \sim N(\bar{\beta}, \bar{V})$ and $\pi(\sigma_\epsilon^2) \sim IG(a, b)$, where IG is the inverse gamma distribution, $\bar{\beta}, \bar{V}$ are the hyper-parameters for β, and a_ϵ, b_ϵ are the hyper-parameters for σ_ϵ^2. For the variance–covariance term Σ_w, we write the conjugate prior $\pi(\Sigma_w) \sim IW(A, \nu)$, that is, inverse Wishart distribution with hyper-parameters A (scale) and ν (degrees of freedom) where $\nu > n-1$. A much simpler version of Σ_w can be written as $\Sigma_w = \sigma_w^2 \mathbf{I}$ which considers the covariance term $\mathbf{S}_w$ to be an identity matrix. In such cases, we can consider an inverse gamma distribution for $\sigma_w^2 \sim IG(a_w, b_w)$ with hyper-parameters a_w (shape) and b_w (scale).

In the above setting of Bayesian modeling, we consider a one-to-one linear function for the true mean process, that is, $\mu = X\beta$, where the predictor X and the mean process μ are not related to the variance or covariance term Σ_w. This leads to defining the distribution of the predictand as $Y \sim N(X\beta, \sigma_\epsilon^2 \mathbf{I})$. In such a case, estimating β is sufficient to explain the model process and we can ignore the estimation for the process μ and its variance σ_w^2 or covariance Σ_w. Hence, we can write

$$\pi(\beta, \sigma_\epsilon^2 | Y) \propto \pi(Y | X, \beta, \sigma_\epsilon^2) \times \pi(\beta, \sigma_\epsilon^2) \tag{2.138}$$

where the conjugate priors $\pi(\beta) \sim N(\bar{\beta}, \bar{V})$, $\pi(\sigma_\epsilon^2) \sim IG(a_\epsilon, b_\epsilon)$ where $\bar{\beta}$ and $\bar{V}$ are the hyper-parameters for β, and a_ϵ, b_ϵ are the hyper-parameters for σ_ϵ^2.

Figure 2.13 shows the Bayesian regression setting states using the directed acyclic graph (DAG) for the process model $\mu = X\beta$ (Koski and Noble, 2011). A DAG is a graphical representation where each edge is directed with an acyclic pattern, that is, the directed graph does not have any feedback loop. The bottom line of Figure 2.13 represents the response variable Y which is linked with the mean process μ and nugget variance σ_ϵ^2. The process μ has the next level of hierarchy where it links with the predictors X and the corresponding parameters β. Then we see the hyper-parameters $\hat{\beta}$ and $\hat{V}$ that link with the parameters β. Similarly, the variance parameter σ_ϵ^2 has hyperparameters a_ϵ and b_ϵ. Gelfand and Banerjee (2017) discuss exact inference in the context of such hierarchical models in the presence of multiple types of data.

2.4.5 Bayesian Model Choice

Bayes' Factor

Let $M = \{M_1, \ldots, M_m\}$ be a set of models with the likelihood for the jth model given by $p_j(y|\theta_j)$, the prior probability of the models being $\pi(M_j)$ and the prior for the jth model parameter being $\pi(\theta_j)$. Using Bayes' theorem, we can write

$$\pi(M_j|y) = \frac{\pi(M_j)p(y|M_j)}{\sum_{l=1}^{m} \pi(M_l)p(y|M_l)}. \tag{2.139}$$

Here $p(y|M_j)$ can be expressed as the marginal likelihood such that

$$p(y|M_j) = \int \pi(\theta_j)p_j(y|\theta_j)d\theta_j. \tag{2.140}$$

The Bayes' factor between a pair of models j and k is defined as follows:

$$\mathrm{BF}_{jk}(y) = \frac{p(y|M_j)}{p(y|M_k)} = \frac{\int \pi(\theta_j)p_j(y|\theta_j)d\theta_j}{\int \pi(\theta_k)p_k(y|\theta_k)d\theta_k}. \tag{2.141}$$

The interpretation is that $\mathrm{BF}_{jk}(y) > 1$ implies that M_j is better than model M_k.

Deviance Information Criterion

The deviance information criterion (Spiegelhalter et al., 2014) is defined as a combination of "goodness of fit" and "complexity" of a model (in a statistical sense; this is different from the idea of "complexity" of a system). We write the expression as

$$\mathrm{DIC} = D(\bar{\theta}) + 2p_D = \overline{D(\theta)} + p_D. \tag{2.142}$$

We will now explain these notations. Let θ be the parameter and its posterior mean $\bar{\theta} = E(\theta|y)$. The term $D(\theta)$ is the Bayesian deviance for θ. The Bayesian deviance can be defined as a relative information measure $-2\log\frac{p(y|\theta)}{p(y|\theta_s)}$ where θ_s is an estimate of θ for a saturated model. Under the Bayesian perspective the parameter θ is random, and hence $D(\theta)$ can be replaced by the posterior expectation $\overline{D(\theta)} = E(D(\theta))$.

The term p_D is the effective number of parameters and is defined by $\overline{D(\theta)} - D(\bar{\theta})$. The p_D can be calculated using simulations of the θ parameters; however, p_D can produce a negative value if the posterior mean of the parameter is far away from the mode (Gelman et al., 2013). Hence, an alternative for p_D can be used, where we define $p_D^{\mathrm{alt}} = 2\,Var\,(\log p(y|\theta))$.

Stochastic Search Selection

Sometimes model choice can be performed by identifying important predictor variables linked to the predictand. This type of method is popular when the number of predictor variables is relatively large. Stochastic search is a method of model selection where a mixture of prior distribution is considered to shrink to zero those model parameters that are not relevant for the model (George and McCulloch, 1993; Ishwaran and Rao, 2005).

Suppose the baseline model is given by $\mathbf{Y} \sim N(\mathbf{X}\beta, \sigma_\epsilon^2 \mathbf{I})$. We define the prior distribution for β_j, $j = 1, \ldots, p$ for each predictor using a latent variable γ_j. In particular, we express the coefficients as products of two random variables in the form of $\beta_j = \gamma_j \Gamma_j$ where $\gamma_j \sim$ Bernoulli(q_j) and $\Gamma_j \sim N(0, \tau_j^2)$. Under this assumption, we can write the prior distribution for β_j as $\pi(\beta_j | \gamma_j) \sim N(0, \gamma_j \tau_j^2)$. The latent variable γ_j can be written as a combination of the spike and slab priors, respectively, where spike refers to a shrinkage in the distribution with a point probability mass (sometimes we use normal distribution with a very small variance and mean zero), and slab includes a relatively wider variance of a normal distribution with mean 0. Notationally, we write $\gamma_j \sim (1 - q_j)\delta_0 + q_j \delta_1$, and δ_0 and δ_1 are the spike and slab priors, respectively. Additionally, we consider $\tau_j^2 \sim IG(a, b)$ (IG denotes inverse gamma distribution) and $q_i \sim$ uniform$(0, 1)$. With respect to the observed data, one can construct the posterior distributions of the parameters to decide on which parameters to keep and which ones to discard. Several versions of stochastic search method have been developed in the literature to define hyper-prior distributions, which can add further sophistication to the variable selection method. See O'Hara and Sillanpää (2009) and Ročková and George (2018) for a comprehensive discussion.

2.5 Multivariate Statistics

Multivariate statistic analysis is useful when we consider more than one variable as the objects of joint analysis (Johnson and Wichern, 2014). It can be based on a dependence method where we simultaneously analyze multiple dependent variables with one or more independent variables. Multivariate regressions, which we will review in this section, belong to this class of analysis. In a complementary fashion, we can define the multivariate case as an interdependence method where we are mainly interested in identifying interrelationship between multiple variables. Naturally, when we deal with only two variables, we refer to it as bivariate analysis.

2.5.1 *Multivariate Normal Distribution*

It is useful to start from the definition of multivariate normal distribution. Let $\mathbf{X} = (X_1, \ldots, X_m)'$ be the $m \times 1$ vector of random variables drawn from such a distribution. We write the m-dimensional normal density function as

$$f_{\mathbf{X}}(\mathbf{x}) = \frac{1}{(2\pi)^{m/2}|\Sigma|^{1/2}} \exp\left[-\frac{1}{2}(\mathbf{x} - \mu)\Sigma^{-1}(\mathbf{x} - \mu)'\right] \qquad (2.143)$$

where μ is a $m \times 1$ vector representing the mean or expectation of $\mathbf{X}$ and Σ is the $m \times m$ variance–covariance matrix of $\mathbf{X}$. Notationally, we describe an m-dimensional normal density by $N_m(\mu, \Sigma)$.

Remark   For $m = 2$, the distribution is referred to as a bivariate normal distribution and we write the variance–covariance matrix as

$$\Sigma = \begin{bmatrix} \sigma_{11} & \sigma_{12} \\ \sigma_{21} & \sigma_{22} \end{bmatrix}$$

where $\sigma_{11} = Var(X_1)$, $\sigma_{22} = Var(X_2)$, and $\sigma_{12} = \sigma_{21} = Cov(X_1, X_2)$. The inverse of the variance–covariance matrix, that is, Σ^{-1} is described as

$$\Sigma^{-1} = \frac{1}{\sigma_{11}\sigma_{22} - \sigma_{12}\sigma_{21}} \begin{bmatrix} \sigma_{22} & -\sigma_{21} \\ -\sigma_{12} & \sigma_{11} \end{bmatrix}.$$

Let us define $\rho_{12} = Corr(X_1, X_2)$ as the correlation between X_1 and X_2. This can be written as

$$\rho_{12} = \frac{\sigma_{12}}{\sqrt{\sigma_{11}}\sqrt{\sigma_{22}}}.$$

Hence, in terms of correlations, we write

$$\Sigma^{-1} = \frac{1}{\sigma_{11}\sigma_{22}(1 - \rho_{12}^2)} \begin{bmatrix} \sigma_{22} & -\rho_{12}\sqrt{\sigma_{11}}\sqrt{\sigma_{22}} \\ -\rho_{12}\sqrt{\sigma_{11}}\sqrt{\sigma_{22}} & \sigma_{11} \end{bmatrix}.$$

Thus, we write the bivariate normal probability density function as

$$f_{X_1,X_2}(x_1, x_2) = \frac{1}{2\pi\sqrt{\sigma_{11}\sigma_{22}(1 - \rho_{12}^2)}} \times$$

$$\exp\left[-\frac{1}{2(1 - \rho_{12}^2)}\left(\left(\frac{x_1 - \mu_1}{\sqrt{\sigma_{11}}}\right)^2 + \left(\frac{x_2 - \mu_2}{\sqrt{\sigma_{22}}}\right)^2\right.\right.$$

$$\left.\left. -2\rho_{12}\left(\frac{x_1 - \mu_1}{\sqrt{\sigma_{11}}}\right)\left(\frac{x_2 - \mu_2}{\sqrt{\sigma_{22}}}\right)\right)\right]. \qquad (2.144)$$

Multivariate Normal Likelihood

Let $\mathbf{X}_1, \ldots, \mathbf{X}_n$ be n samples randomly drawn from an m-dimensional multivariate normal distribution with mean μ and variance–covariance matrix Σ. Let the draws be independent with each following a normal distribution: $\mathbf{X}_1 \sim N_m(\mu, \Sigma)$, $\ldots$, $\mathbf{X}_n \sim N_m(\mu, \Sigma)$. In such a scenario, we can write the joint likelihood function as

$$L(\mu, \Sigma) = \prod_{i=1}^{n} \left[\frac{1}{(2\pi)^{m/2}|\Sigma|^{1/2}} \exp\left[-\frac{1}{2}(\mathbf{x}_i - \mu)\Sigma^{-1}(\mathbf{x}_i - \mu)'\right]\right]$$

$$= \frac{1}{(2\pi)^{mn/2}|\Sigma|^{n/2}} \exp\left[-\frac{1}{2}\sum_{i=1}^{n}(\mathbf{x}_i - \mu)\Sigma^{-1}(\mathbf{x}_i - \mu)'\right]. \quad (2.145)$$

Remark The maximum likelihood estimators of mean μ and covariance matrix Σ are defined as $\hat{\mu} = \bar{\mathbf{x}}$ and $\hat{\Sigma} = \frac{1}{n}\sum_{i=1}^{n}(\mathbf{x}_i - \bar{\mathbf{x}})(\mathbf{x}_i - \bar{\mathbf{x}})'$. From a sampling perspective, we write the sample variance–covariance matrix as $\mathbf{S} = \frac{1}{n-1}\sum_{i=1}^{n}(\mathbf{x}_i - \bar{\mathbf{x}})(\mathbf{x}_i - \bar{\mathbf{x}})'$. Therefore, we can write $\hat{\Sigma} = \frac{n-1}{n}\mathbf{S}$. Here $\bar{\mathbf{x}}$ and $\mathbf{S}$ are sufficient statistics for μ and Σ.

Consider the same $\mathbf{X}_1, \ldots, \mathbf{X}_n$ as above with $\bar{\mathbf{x}}$ and $\mathbf{S}$ being the corresponding sufficient statistics; $\bar{\mathbf{x}}$ is normally distributed with mean μ and covariance matrix $(1/n)\Sigma$. The distribution of the sample covariance matrix is obtained from the sum of products of $\mathbf{X}$ and known as the Wishart distribution. We write the Wishart distribution with m degrees of freedom as $W_m(\Sigma)$ which follows the distribution of $\sum_{j=1}^{m}\mathbf{x}_j\mathbf{x}_j'$.

2.5.2 Multivariate Hypothesis Testing

Hotelling T^2

Often under a multivariate setting, the goal is to test a hypothesis for a set of mean values $\mu = (\mu_1, \ldots, \mu_n)'$. Suppose we want to test the null hypothesis $H_0 : \mu = \mu_0$ with an alternative $H_1 : \mu \neq \mu_0$. We denote the test statistic as Hotelling T^2 and define it as

$$T^2 = n(\bar{\mathbf{x}} - \mu_0)'\mathbf{S}^{-1}(\bar{\mathbf{x}} - \mu_0) \quad (2.146)$$

where $\bar{\mathbf{x}} = (1/n)\sum_{i=1}^{n}\mathbf{x}_i$ is an $n \times 1$ vector and $\mathbf{S} = \frac{1}{n-1}\sum_{i=1}^{n}(\mathbf{x}_i - \bar{\mathbf{x}})(\mathbf{x}_i - \bar{\mathbf{x}})'$. The Hotelling T^2 is distributed as

$$T^2 \sim \frac{(n-1)m}{n-m}F_{m,n-m} \quad (2.147)$$

where $F_{m,n-m}$ is an F-distribution with m and $n - m$ degrees of freedom. For hypothesis testing, let α be the level of significance; we reject H_0 at level α when

$$T^2 > \frac{(n-1)m}{n-m}F_{m,n-m}(\alpha). \quad (2.148)$$

Wilks' Lambda

Building further on our discussion on the likelihood-ratio method, we denote the test statistic as Wilks' lambda in multivariate settings. Suppose the hypothesis is

$H_0 : \mu = \mu_0$ and under H_0, we denote the likelihood function by $L(\mu_0, \Sigma)$. The likelihood ratio is written as

$$\Lambda = \frac{\max_{\Sigma} L(\mu_0, \Sigma)}{\max_{\mu, \Sigma} L(\mu, \Sigma)} = \left[\frac{|\hat{\Sigma}|}{|\hat{\Sigma}_0|} \right]^{n/2} \tag{2.149}$$

where $|M|$ denotes determinant of a matrix M. A slight modification yields the test statistic, Wilks' lambda, with the following expression:

$$\Lambda^{2/n} = \left[\frac{|\hat{\Sigma}|}{|\hat{\Sigma}_0|} \right]. \tag{2.150}$$

We can represent the Wilks' lambda in terms of Hotelling T^2 statistic as well:

$$\Lambda^{2/n} = \left[1 + \frac{T^2}{n-1} \right]^{-1}. \tag{2.151}$$

To test against $H_1 : \mu \neq \mu_0$, we reject H_0 for a small value of $\Lambda^{2/n}$ or a large value of T^2, with the critical region being defined for the Hotelling T^2 test statistic.

Multivariate Analysis of Variance

The multivariate analysis of variance (MANOVA) is a multivariate version of ANOVA (which we discussed in Section 2.3.3), where we compare several multivariate means. It is used to jointly analyze multiple dependent variables. Suppose we have random samples from K populations, where each population follows a multivariate normal distribution. Let $\mathbf{X}_{k1}, \ldots, \mathbf{X}_{kn_k}$ be a random sample from a multivariate normal population with mean μ_k, where $k = 1, \ldots, K$. We assume that the populations are independent and have a common variance Σ. We write the model as follows:

$$\mathbf{X}_{kj} = \bar{\mathbf{X}} + \alpha_k + \epsilon_{kj}, \quad j = 1, \ldots, n_k; \quad k = 1, \ldots, K. \tag{2.152}$$

Here, let $\bar{\mathbf{X}}$ be the overall or grand sample mean; $\alpha_k = (\bar{\mathbf{X}}_k - \bar{\mathbf{X}})$ denotes the deviation of the kth group mean such that $\sum_{k=1}^{K} n_k \alpha_k = \mathbf{0}$. The error term $\epsilon_{kj} = (\mathbf{X}_{kj} - \bar{\mathbf{X}}_k)$ is assumed to follow a normal distribution: $\epsilon_{kj} \sim N(\mathbf{0}, \Sigma)$. The null hypothesis boils down to the following:

$$H_0 : \alpha_1 = \alpha_2 = \cdots = \alpha_K = 0.$$

Similar to ANOVA, in MANOVA we construct between ($\mathbf{B}$) and within ($\mathbf{W}$) sums of squares with the following expressions:

$$\mathbf{B} = \sum_{k=1}^{K} n_k (\bar{\mathbf{X}}_k - \bar{\mathbf{X}})(\bar{\mathbf{X}}_k - \bar{\mathbf{X}})'$$

$$\mathbf{W} = \sum_{k=1}^{K} \sum_{j=1}^{n_k} (\mathbf{X}_{kj} - \bar{\mathbf{X}}_k)(\mathbf{X}_{kj} - \bar{\mathbf{X}}_k)'$$

where $\mathbf{B}$ and $\mathbf{W}$ have $K - 1$ and $\sum_{k=1}^{K} n_k - K$ degrees of freedom, respectively. Hence, we reject H_0 if the ratio

$$\Lambda^* = \frac{|\mathbf{W}|}{|\mathbf{W} + \mathbf{B}|}$$

is very small. For large samples, the statistic

$$-\left(n - 1 - \frac{p+K}{2}\right) \log \Lambda^* = -\left(n - 1 - \frac{p+K}{2}\right) \log \frac{|\mathbf{W}|}{|\mathbf{W} + \mathbf{B}|} \quad (2.153)$$

approximately follows a chi-square distribution with $p(K - 1)$ degrees of freedom, where p is the number of variables and $\sum_{k=1}^{K} n_k = n$. Hence, we reject H_0 at α level of significance if

$$-\left(n - 1 - \frac{p+K}{2}\right) \log \frac{|\mathbf{W}|}{|\mathbf{W} + \mathbf{B}|} > \chi^2_{p(K-1)}(\alpha) \quad (2.154)$$

where $\chi^2_{p(K-1)}$ is the chi-square distribution with $p(K - 1)$ degrees of freedom.

Remark MANOVA is sensitive to data outliers. Additionally, if the dependent variables are a linear combination of each other, that is, they are highly correlated, then this technique would not be useful.

2.5.3 Multivariate Methods

Dependence Method

In a dependence multivariate model, the model has at least two dependent or response variables which depend on one or more independent or predictor variables. In case of exactly two dependent variables, we denote the model as a bivariate model. As in the univariate regression problem, we establish a relationship between the dependent and independent variables.

Multivariate regression: Let $\mathbf{Y}$ be the $n \times m$ matrix of dependent variables which depends on the independent variable matrix $\mathbf{X}$ of order $n \times p$ with model parameters β matrix of order $p \times m$. Thus we write the multivariate multiple regression model

$$\underset{n \times m}{\mathbf{Y}} = \underset{n \times p}{\mathbf{X}} \times \underset{p \times m}{\beta} + \underset{n \times m}{\epsilon} \quad (2.155)$$

where we denote the matrices of the multivariate regression model as

$$\underset{n \times m}{\mathbf{Y}} = \left[\mathbf{Y}_{n \times 1}^{(1)}, \ldots, \mathbf{Y}_{n \times 1}^{(m)}\right], \quad \underset{n \times p}{\mathbf{X}} = \left[\mathbf{X}_{n \times 1}^{(1)}, \ldots, \mathbf{X}_{n \times 1}^{(p)}\right]$$

$$\underset{p\times m}{\beta} = \left[\beta_{p\times 1}^{(1)}, \ldots, \beta_{p\times 1}^{(m)}\right], \quad \underset{n\times m}{\epsilon} = \left[\epsilon_{n\times 1}^{(1)}, \ldots, \epsilon_{n\times 1}^{(m)}\right]$$

where $\underset{n\times m}{\epsilon} \sim N(\mathbf{0}_{n\times m}, \Sigma \otimes \mathbf{I})$ and $\otimes$ denotes a Kronecker product. The least square estimates of $\hat{\beta}_{p\times 1}^{(l)}, l = 1, \ldots, m$ for the multivariate regression are computed individually for each response variable. Thus, $\hat{\beta}_{p\times 1}^{(l)} = (\mathbf{X}'\mathbf{X})^{-1}\mathbf{X}'\mathbf{Y}^{(l)}$ for $l = 1, \ldots, m$. We can also develop Bayesian versions of multivariate regression by assigning multivariate prior distributions to the model parameters.

Interdependence Method

Unlike in dependence methods, in an interdependence method we consider two or more variables on equal footing in terms of their relationship. This is a predominantly exploratory approach to identify similarities between the variables, corresponding groupings, and categorization. Two important techniques are principal component analysis and clustering analysis. In principal component analysis (PCA), the goal is to explain a large fraction of the overall variance of the variables with a few recreated composite variables. This helps to reduce a large number of variables into a small number of composite variables. In clustering analysis, we identify groups based on characteristics of variables, which are similar in nature within groups and dissimilar to other groups. The principal components can also be used for clustering. However, there are several other clustering methods available for grouping based on variable characteristics which are more intricate than PCA and yield different insights. We will not discuss these here but will explore these two topics in detail in Chapter 4 when we discuss machine learning techniques.

2.6 Taking Stock and Further Reading

This chapter has provided an overview of a few commonly used statistical tools. We started our discussion by explaining data types and the fundamentals of population and sample. Next we introduced the concepts of probability, distributions, and moments, which are essential building blocks of statistics. The discussion then led to classical inference, which includes estimation process and modeling strategy. For a more elaborate treatment of this material, one can consult Hogg et al. (2005) and Rice (2006).

Next, we discussed Bayesian inference following the rationale of the Bayesian approach. Our discussion was confined within the conceptual framework of Bayesian statistics. There are multiple topics that we did not discuss. For an in-depth review, interested readers can refer to Gelman et al. (2004) who provide a very comprehensive treatment of Bayesian analysis. An extension of Bayesian statistics comes in the form of Bayesian networks as they relate to machine learning.

These are directed acyclic graphs with nodes that represent stochastic variables, used for prediction and inference (Chen and Pollino, 2012). Finally, we briefly discussed ideas related to multivariate analysis and ways of thinking about statistical distributions in a multivariate setup. Chatfield and Collins (1981) provide an elaborate treatment of multivariate statistics. A more up-to-date and applied exposition can be found in Everitt and Hothorn (2011). We will utilize many of the concepts described and discussed above in the following chapters.

As indicated above, the coverage of topics here is far from being exhaustive. Here we provide some additional references for background material and new developments that interested readers may consult. In our discussion, we have completely skipped discussing probability in a rigorous way. A few classic references on probability are Feller (2008a), Feller (2008b), and Billingsley (2008).

In recent times, statistical methodology has branched out in many different dimensions and significant developments have happened in interdisciplinary fields. Here we refer to only two such literatures. The first one is about causality, where econometrics has also made big contributions (Angrist and Pischke, 2008). In fact, the Nobel Prize for economics in 2021 was given to Joshua Angrist and Guido Imbens in recognition of their contributions to the analytical understanding of causality. Historically, Yule (1900) presented probably one of the first treatments of causality in the form of association. The conceptual understanding of causality and association developed significantly over the last century. D. R. Cox commented on two views of statistical causality that one view "is a statistical association that cannot be explained away by confounding variables and the other is based on a link with notions in the design of experiments" (Cox, 1992). Later Judea Pearl provided a systematic treatment on the topic and a detailed analysis of causality from a complementary perspective (Pearl, 2009). A review of different ways of viewing statistical causality can be found in Cox and Wermuth (2004) (see also Altman and Krzywinski, 2015). But by and large, statistics literature has not historically been very prominent in tackling the question of causality (see the first chapter of Imbens and Rubin [2015] for a discussion on this). The recent literature is significantly more active in this domain. Imbens and Rubin (2015) provide a review of the frontier literature in this field and present probably one of the best expositions on this topic.

The second body of literature we refer the reader to is about *big data*. For analyzing such data, Efron and Hastie (2016) present an intriguing analysis of statistical inference through a computational approach (see also Lafferty and Wasserman, 2006). A major complementary development has taken place at the intersection of statistics and machine learning literature. For readers interested in the statistical treatment of machine learning techniques, a very good resource is James et al. (2013) (see also Hastie et al., 2009). Their approach complements our discussion on machine learning techniques in the later part of this book.

3

Statistical Analyses of Time-Varying Phenomena

In this chapter, we will utilize the concepts discussed in the previous chapter on statistics and probability, and describe a toolkit to analyze time-varying behavior of real-world systems. Real-world systems are messy and predictions may not be exactly trustworthy. Economists know this very well. In fact, the repeated failure of precise forecasting ability led to the adage that an economist is an expert who can figure out in future why the predictions that were made in the past are not taking place in the present! Although this description was originally about economics, it is probably applicable for complex systems as well, especially given the fact that one of the defining features of complex systems is the emergence of features that may not be traced back to the behavior of an average constituent element. However, we will see that one can still make a lot of sense of time-varying phenomena from observational data.

There are abundant examples of systems that generate time-varying responses. Rainfall, temperature, gross domestic production in a region, all of these variables fluctuate over time. Some of them fluctuate in a manner more periodical than others: for example, rainfall may peak around July in eastern India whereas snowfall may peak around January on the east coast of the USA. Over 24 h one might keep track of the number of cars passing by in a busy street in a city. A likely pattern would be that the number of cars passing by would increase between 9 a.m. and 12 noon and again between 5 and 10 p.m. In contrast, stock price returns seem to exhibit almost no patterns of periodic oscillation.

Any such observation can be represented by a time series object, which can be simply thought of as a vector of values recorded over time. To fix the notations, we denote a time series by $X_T = \{x_1, x_2, \ldots, x_T\}$. Here x_t denotes an observation (say, rainfall) recorded at time t where the time index t varies between 1 and T, T being a positive integer. For all t, the values would be real: that is, $x_t \in \mathbb{R}$ (we will not consider complex values).

Here, we present two examples of time series from two very different systems. In Figure 3.1, we show the evolution of total global production measured by world

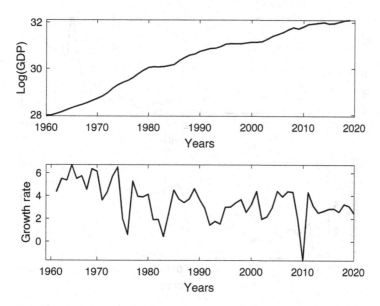

Figure 3.1 Total global production over six decades. Upper panel: Evolution of world gross domestic product (GDP) from 1961 to 2020. Lower panel: Corresponding growth rates (source: World Bank).

gross domestic product from 1960 to 2020 (data publicly available from the World Bank website). The top panel shows the increasing trend in total production on a logarithmic scale. In the bottom panel, we have plotted the yearly growth rates of the same series, which exhibits a lot more fluctuations. For example, we see that around the time of the global financial crisis (2008–9) global growth turned negative. Another example of a time series can be seen in the top panel of Figure 3.2 which presents the evolution of the market capitalization of Bitcoin. The data clearly shows an initial period of slow growth with mild fluctuations followed by a period of explosive growth in market size. However, in the log scale (bottom panel), the growth seems to have a linear component with a clear upward trend and occasional fluctuations.

In this chapter, our task will be twofold. First, we will create models that will allow us to see if there is any connection in a statistical sense between values observed over time for a given system. Second, we will discuss and describe statistical apparatus to estimate time series models based on a given set of observations. The models and techniques we will see in this chapter can be directly applied to many different types of time series data: for example, from finance, physics, or climate science. However, for the sake of exposition, we note that many of these techniques were specifically built for economic and financial variables. Therefore, our approach will also be from that angle.

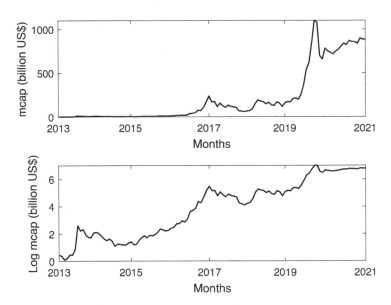

Figure 3.2 Evolution of market capitalization of Bitcoin from April 2013 to August 2021 (based on publicly available online data). Upper panel: In level. Lower panel: In logarithm.

3.1 Some Basic Definitions and Constructions

Let us start with a time series object: $X_T = \{x_1, x_2, \ldots, x_T\}$. We will interchangeably use X_T or $\{x_t\}$ to denote the time series. For the time being, we will assume that X_T is obtained from a *stationary* process. Intuitively, what we mean is that such processes are *well-behaved* and have nice statistical properties. We will elaborate later on exactly what we mean by "stationarity" in a more technical sense.

Let us first define the mean of our observations as $\mu(X_T) = 1/T \sum_{t=1}^{T} x_t$ and the corresponding variance as $\sigma^2(X_T) = 1/T \sum_{t=1}^{T} (x_t - \mu)^2$. The population analogues will be denoted below by $E(.)$ and $Var(.)$. Because these observations are recorded over time t, a starting point to see if there is any pattern would be to check if a high value of x is followed by a high value of x or if consecutive values have no relationship whatsoever. To quantify this idea, we define *autocovariance*, which allows to see if there are any lagged comovements in the recorded observations. For notational purposes, we will say that observation x_{t-j} lags observation x_t by j lags. Formally, autocovariance at the jth lag is given by

$$\gamma_j = Cov(x_t, x_{t-j}). \tag{3.1}$$

For stationary processes, the time index t doesn't matter as covariance across j time points will be the same for all time points t. If the process has a zero mean, then the j-lag autocovariance can be simply written as

$$\gamma_j = E(x_t x_{t-j}). \tag{3.2}$$

Note that 0-lag autocovariance is just the variance:

$$\gamma_0 = Var(x_t). \tag{3.3}$$

Typically it is more useful to scale the autocovariances by the variance so that different series with different degree of variances can be compared. The normalized values represent *autocorrelations*.

Definition 3.1   The autocorrelation function (ACF) is defined as

$$\rho_j = \frac{\gamma_j}{\gamma_0} \tag{3.4}$$

for the *j*th lag (*j* can be positive or negative).

3.1.1 Frequency and Time Domains

Although we will deal almost exclusively with time domain analysis of time series objects, it will be useful here to provide a brief introduction to frequency domain representation. For that purpose, we need to first define two terms: *period* and *frequency*. The period of a wave is the time τ that a wave takes to go through a whole cycle (2π rotation). The frequency ν is the number of whole cycles completed in unit time by a wave.

Combining these two definitions we can write

$$\tau = \frac{2\pi}{\nu}. \tag{3.5}$$

Consider the time interval to be discrete. There exists a very useful tool that allows us to go back and forth between the time domain and the frequency domain. Formally it is called the *Fourier transform* after its inventor Joseph Fourier. For a time series $\{x_t\}$, we can write its Fourier transformation as

$$x(w) = \frac{1}{2\pi} \sum_{t=-\infty}^{\infty} e^{-it\nu} x(t). \tag{3.6}$$

We can also carry out the reverse operation in the form of an inverse Fourier transform:

$$x(t) = \int_{-\pi}^{\pi} e^{itw} x(\nu) d\nu. \tag{3.7}$$

We will see that instead of converting the series itself into frequencies, it would be more useful to deal with the autocovariance function and to study its frequency spectrum.

3.1.2 Autocovariance Function

Before we get into the derivations in this section, a small point on notations is in order. Here, we will deal with complex numbers as well as lags in time series. We will continue to denote the complex number $\sqrt{-1}$ by i (as we did when describing characteristic functions in Section 2.2.3 of Chapter 2) and the lags by h. We will not use j here in the way that we did in the previous section to denote the lags. The reason is that often j is also used to denote complex numbers. Unless we explicitly state otherwise, we will use this convention of writing complex numbers only within this section (Section 3.1.2). Also, the following discussion will assume that the time index of the series goes from negative to positive infinity.

The following description and derivations have been partially inspired by DeJong and Dave (2011). As we described in the previous section, the autocovariance function for a zero-mean stationary process $\{x_t\}$ is given by

$$\gamma_h = E(x_t x_{t-h}). \tag{3.8}$$

We define the spectrum of the time series $\{x_t\}$ as

$$S_x(v) = \frac{1}{2\pi} \sum_{h=-\infty}^{\infty} e^{-ihv} \gamma_h. \tag{3.9}$$

This is a useful expression. Let us evaluate the expression at $v = 0$ (Equation 3.9). A direct substitution gives us

$$\sum_{h=-\infty}^{\infty} \gamma_h = 2\pi S_x(0). \tag{3.10}$$

Thus, the strength of the spectrum evaluated at $v = 0$ is proportional to the sum of all autocovariances. We can simplify the right-hand side of Equation 3.9 by using Euler's identity:

$$e^{i\theta} = \cos\theta + i\sin\theta. \tag{3.11}$$

Simply by substituting the exponential terms e^{-ihv} from the Euler's identity, the equation boils down to

$$S_x(v) = \frac{1}{2\pi}\left[\gamma_0 + 2\sum_{h=1}^{\infty} \gamma_h \cos(hv)\right]. \tag{3.12}$$

Here we have used the identity that $\cos(-v) = \cos(v)$ and $\sin(-v) = \sin(v)$.

We can also use the inverse Fourier transform to generate autocovariance in a reverse operation:

$$\gamma_h = \int_{-\pi}^{\pi} e^{ivh} S_x(v) dv. \tag{3.13}$$

We can make an alternate representation by utilizing the formalism of moment-generating functions that we introduced in Chapter 2. The autocovariance-generating function can be written as

$$g_x(z) = \sum_{h=-\infty}^{\infty} \gamma_h z^h. \tag{3.14}$$

Therefore, if we carry out a substitution $z = exp(-iv)$, then the spectrum is simply given by

$$S_x(v) = \frac{g_x(z)}{2\pi}. \tag{3.15}$$

Before ending the present discussion, we note that one can carry out all the above analysis in terms of the autocorrelation function as well by dividing both sides by $\gamma_x(0)$. The analog of Equation 3.9 would be

$$f(v) = \frac{1}{2\pi} \sum_{h=-\infty}^{\infty} e^{-ihv} \rho_h. \tag{3.16}$$

With the inverse Fourier transform, we can recover the autocorrelation function as

$$\rho_h = \int_{-\pi}^{\pi} e^{ivh} f_x(v) dv. \tag{3.17}$$

This approach is very useful in the context of designing filters to extract specific signal contents from an observed time series. Readers interested in derivations of such filters and their applications in macroeconomics can refer to DeJong and Dave (2011).

The above description shows that it is possible to go back and forth between frequency domain representation and time domain representation for a time series. The usefulness of frequency domain representation becomes very clear when analyzing signals for their frequency contents. However, in the present context we will not focus on regular oscillatory phenomena which have clear spikes in frequencies, making the periodicity in the time series prominent. We will instead focus on time series for which the periodic behavior is less clear and although oscillations exist, they are not as regular as those of periodic waves. Often the time domain analysis for such time series is easier to conduct and interpret.

At this stage, it will be useful to consider some examples of variables for which this kind of modeling would be appropriate. An useful example of the type of data that we can analyze with time series modeling would be *business cycles*. These are defined as fluctuations found in aggregate economic activities in a given country, typically calculated as the deviations of per capita gross domestic product from its long-term growth path. The crests are called *booms* and the troughs are called *recessions*. The important point to note here is that these are not regular cycles with given frequencies. Economists and policy-makers have been busy predicting the timing of these economic booms and recessions for a very long time, with mixed success. The lack of success provides clear evidence of the fact that such oscillations are not periodic in nature. Other examples would be asset prices (such as stocks and bonds) or, in the context of physical world, the modeling of temperature, rainfall, and so forth.

3.2   Stationary Time Series

In this section, we will analyze well-behaved time series, roughly corresponding to time series that do not exhibit explosive tendency over time and whose underlying data-generating process does not change over time. However, we need a formal definition of exactly what we mean by *well-behaved*. The way to capture the idea is to consider the concept of *stationarity*.

3.2.1   What Is Stationarity?

There are two types of stationarity. First, we consider the strong version of stationarity. We call a time series strongly stationary if the joint distribution of $x_{t_1}, \ldots, x_{t_k}$ remains identical with respect to the joint distribution of the variables shifted over τ time points: $x_{t_1+\tau}, \ldots, x_{t_k+\tau}$ for all k and τ.

The properties of *strong stationarity* are too strong to be useful in our current discussion, as it requires the joint probability density function to be preserved. For our purpose, a weaker type of stationarity suffices. The following version is called a weakly stationary process. A weakly stationary stochastic process is defined as a process for which the mean $E(x_t) = c$ where c is a constant that is independent of time t; the variance $Var(x_t) = \sigma^2$ where σ is independent of time t; and finally, the covariance $Cov(x_t, x_{t-j}) = \gamma_j$; that is, it depends only on the lag j. Sometimes a weakly stationary process is also called a *covariance stationary* process.

It is worthwhile to note that a strongly stationary series is not necessarily weakly stationary (e.g. may not have a finite second moment) and vice versa (e.g. higher-order moments can be time-varying).

3.2.2   Building a Time Series Model

So far we have described and characterized different properties of a time series vector. Now, we would like to build a model that can mimic the properties of a given time series $\{x_t\}$. The building block of time series models is the *white noise* process.

Let us denote a white noise process by the variable ε_t. The idea behind a white noise process is that the distribution of ε_t does not change over time and an outcome of ε_t does not affect the probability of any future outcomes. In short, a simple representation of white noise would be an independent and identically distributed (IID) variable. Formally, a *white noise* process would exhibit zero serial correlation and homoscedasticity. In order to work with ε_t, we impose some additional desirable properties. Generally, we assume that the variable is normalized to have zero mean. Before we get into the technical discussion of white noise, here is a fun piece of trivia. One may ask – why is white noise called "white"? The answer is that if one constructs the spectrum of color following the discussion in Section 3.1.2 above, then it can be shown that all frequencies would be present with equal strength, a property which resembles white color. That is why the process is called white noise.

A simple and very commonplace example of white noise is independently distributed normal random variables:

$$\varepsilon_t \sim \text{IID } N(0, \sigma_\varepsilon^2). \tag{3.18}$$

Then, the implications of the abovementioned assumption would be: for all t, $E(\varepsilon_t) = 0$, $Cov(\varepsilon_t, \varepsilon_{t-j}) = 0$ for all j, and $Var(\varepsilon_t) = \sigma_\varepsilon^2$.

Given this simple stochastic process, we can build a wide range of time series models simply by linearly combining the building blocks: that is, ε_ts. But before discussing the model-building exercise, we need to establish a crucially important result regarding the scope of all the models we can build. Suppose we can build only a small set of models with our building blocks, the ε_ts. Then if we observe a time series that falls outside this class of models, we will have to look for newer toolkits and modeling methodologies. Fortunately for us, there exists a very useful and influential result called the *Wold decomposition theorem* which basically states that any weakly stationary process would be amenable to modeling via a linear combination of white noise. Thus, as long as we can ensure weak stationarity of the process, we are assured to have a model for it.

Theorem 3.2 (Wold decomposition theorem (Prop. 4.1, Hamilton, 1994)). Any mean zero covariance stationary process $\{x_t\}$ can be represented as the sum of one non-deterministic component and one linearly deterministic component. Mathematically, one can write the decomposition as

$$x_t = \sum_{j=0}^{\infty} \beta_j \varepsilon_{t-j} + \theta_t \tag{3.19}$$

"where ε_t represents white noise, $\beta_0 = 1$, the possibly infinite sequence β_j is square summable (i.e. $\sum_{j=1}^{\infty} \beta_j^2 < \infty$), and θ_t is deterministic in the sense that a linear function defined over the past values of $\{x_t\}$ can predict θ_t arbitrarily well."

The tremendous importance of this theorem arises from the fact that any weakly stationary process has this representation in terms of linear combination of white noise. This is a very useful feature that we will utilize below when building time series models.

3.2.3 The ARMA Class of Models

AutoRegressive Moving Average (ARMA) models are created by literally taking linear combinations of white noise. We will first set up the model and then describe a succinct representation of the model in terms of lag polynomials, which make it much easier to analyze the time series in terms of their correlation structure. Then we will extend the system of equations to consider multiple variables at once. Readers can refer to Hamilton (1994) for a very detailed and analytical introduction to ARMA models.

The simplest model of this class, an autoregressive process of order 1, in short AR(1), is given by

$$x_t = \alpha x_{t-1} + \varepsilon_t \tag{3.20}$$

where α is a constant, often called the AR coefficient. To get a sense of the process, let us say that time starts from $t = 0$ when the value of the variable x is x_0. Then we can recursively construct the sequence by adding white noise as follows:

$$x_1 = \alpha x_0 + \varepsilon_1$$
$$x_2 = \alpha x_1 + \varepsilon_2$$
$$\vdots \tag{3.21}$$

and so forth. A different but complementary class of models is produced by moving average (MA) processes. An MA(1) process is defined as follows:

$$x_t = \beta \varepsilon_{t-1} + \varepsilon_t. \tag{3.22}$$

These processes can easily be generalized to multiple lags. We can define an AR(p) process as follows:

$$x_t = \sum_{i=1}^{p} \alpha_i x_{t-i} + \varepsilon_t, \tag{3.23}$$

and an MA(q) process as follows:

$$x_t = \sum_{j=0}^{q} \beta_j \varepsilon_{t-j}. \tag{3.24}$$

By combining the AR and the MA terms, we get the general form ARMA(p, q):

$$x_t = \alpha_0 + \sum_{i=1}^{p} \alpha_i x_{t-i} + \sum_{j=1}^{q} \beta_j \varepsilon_{t-j} + \varepsilon_t \tag{3.25}$$

where α_0 is a constant and β_0 is normalized to 1. Without loss of generality, we assume that $E(x) = 0$: that is, the process has a zero mean.

Note that we have not specified the range of values that the parameter vectors $\{\alpha_i\}$ and $\{\beta_j\}$ can take. Intuitively it is obvious that they are not unrestricted. For example, consider the AR(1) process described in Equation 3.20. If the coefficient $\alpha > 1$, then the sequence of values of x will explode. Therefore, it will not be a stationary series any more (recall that for weak stationarity, the series must possess a finite second moment along with satisfying other conditions). Thus, we need some restrictions on the parameter values to make sure that the processes remain stationary. The standard way to do this is to consider a representation in terms of lag polynomials.

Representation in Terms of Lag Polynomials

First, we have to define lag or backshift operator. This operator simply allows us to go back and forth on a time series by advancing and retreating in time. We will interchangeably use *lag* and *backshift* operator, and use the notation $\mathcal{B}$ to denote it. Some textbooks use the notation L to denote the same (e.g. Hamilton, 1994). Formally, we write

$$\mathcal{B}(x_t) = x_{t-1}. \tag{3.26}$$

While in principle we can also work with its inverse lead operator

$$\mathcal{B}^{-1}(x_{t-1}) = x_t, \tag{3.27}$$

usually it is easier to describe it in the form of lags. This operator provides a very useful and succinct way to deal with the parameter vectors $\{\alpha_i\}$ and $\{\beta_j\}$ along with the lagged values of x_t and the white noise term ε_t. We can define a polynomial over the lag operators in the following way:

$$\alpha(\mathcal{B}) = \sum_{i=0}^{n} \alpha_i \mathcal{B}^i \quad \text{and} \quad \beta(\mathcal{B}) = \sum_{j=0}^{n} \beta_j \mathcal{B}^j. \tag{3.28}$$

In this notation, the AR(1) process can be written $(1 - \alpha \mathcal{B}) x_t = \varepsilon_t$ and MA(1) process can be written as $x_t = (1 + \beta \mathcal{B}) \varepsilon_t$, where we have ignored the subscripts α and β since there are only single instances of them in each process. The generalized version for an ARMA(p, q) process can be written concisely as

$$\alpha(\mathcal{B}) x_t = \beta(\mathcal{B}) \varepsilon_t. \tag{3.29}$$

As will be evident in the following text, this representation in terms of lag polynomials is very useful for two purposes. First, it allows us to interchange between the

AR and MA representations, and second, the invertibility of lag polynomials gives us the conditions for the existence of a stationary process.

Let us first elaborate on the ease of going from one representation to the other. Let's consider the process from Equation 3.20:

$$x_t = \alpha x_{t-1} + \varepsilon_t \quad \text{where } |\alpha| < 1. \tag{3.30}$$

The condition $|\alpha| < 1$ is required to ensure stationarity of the process. For real data, this parameter needs to be estimated. For the time being, let us assume that this condition is satisfied. The easiest way to think about mapping x_t into the values of the white noise term ε_t is to simply recursively substitute terms:

$$
\begin{aligned}
x_t &= \alpha x_{t-1} + \varepsilon_t \\
&= \alpha^2 x_{t-2} + \alpha \varepsilon_{t-1} + \varepsilon_t \\
&\vdots \\
&= \alpha^t x_0 + \sum_{j=0}^{t-1} \alpha^j \varepsilon_{t-j}.
\end{aligned} \tag{3.31}
$$

Note that here we already see why $|\alpha| < 1$. Specifically, under that condition, if we assume that the process started infinite periods ago, we have

$$x_t = \sum_{j=0}^{\infty} \alpha^j \varepsilon_{t-j}. \tag{3.32}$$

There is an alternative method to get the same expression. Using the lag operator, we know that the AR(1) process can be written as

$$(1 - \alpha B)x_t = \varepsilon_t. \tag{3.33}$$

We can expand on the expression (assuming $|\alpha| < 1$):

$$
\begin{aligned}
x_t &= \frac{\varepsilon_t}{(1 - \alpha B)} \\
&= (1 + \alpha B + \alpha^2 B^2 + \alpha^3 B^3 + \cdots)\varepsilon_t \\
&= \varepsilon_t + \alpha \varepsilon_{t-1} + \alpha^2 \varepsilon_{t-2} + \cdots \\
&= \sum_{j=0}^{\infty} \alpha^j \varepsilon_{t-j}.
\end{aligned} \tag{3.34}
$$

Following the same logic, we can expand the AR(p) process to MA(∞) as long as the lag polynomials are invertible. To see why invertibility is important, recall from Equation 3.25 that a standard ARMA(p, q) model can be written as

$$\alpha(B)x_t = \beta(B)\varepsilon_t. \tag{3.35}$$

Assuming that the lag polynomials $\alpha(B)$ and $\beta(B)$ are invertible, we write the same process in two different fashions:

$$x_t = \alpha(B)^{-1}\beta(B)\varepsilon_t,$$
$$\varepsilon_t = \beta(B)^{-1}\alpha(B)x_t. \tag{3.36}$$

Note that in the above calculations (e.g. in Equation 3.34), we have treated the lag polynomial as a regular algebraic polynomial and conducted the Taylor expansion in the usual way. One question naturally arises as to whether this is a permissible operation or not. The short answer is: yes, it is a permissible operation. We will not elaborate on the reason here. Interested readers can consult Hamilton (1994) for a detailed description covering, in particular, the fact that the lag operator satisfies commutative, associative, and distributive laws for multiplication and addition.

3.2.4 Generalization to the Vector Autoregression Model

So far we have dealt with only one time series $\{x_t\}$. In many cases, multiple time series might naturally coexist. For example, one can consider asset returns in a multi-asset market or the growth rates of different firms in an economy. The interesting feature of such a scenario is that not only do the variables have non-trivial time-dependence, they also affect each other. Thus there may exist mutual interaction across the variables. The *vector autoregression* (VAR) model is a generalized version of the simple autoregression model that allows us to capture such mutual dependence across variables. It is worthwhile to note that, in principle, one can describe a *vector autoregressive moving average* model as well. However, typically a well-specified VAR model provides enough flexibility to capture the patterns in the data and it is less complicated than a VAR model augmented with moving average terms. Also, a VAR model can be written as a combination of infinite number of moving average terms. Therefore, we focus only on the VAR setup.

Let us begin with a three-variable example with self- and cross-dependence on only one lag. Let us imagine we have three variables x_{1t}, x_{2t}, and x_{3t} that are related to each other's lagged values linearly and that they themselves also depend on their own past values. In vector notation, the variables and the corresponding white noise terms can be written as

$$x_t = \begin{pmatrix} x_{1t} \\ x_{2t} \\ x_{3t} \end{pmatrix}, \qquad \varepsilon_t = \begin{pmatrix} \varepsilon_{1t} \\ \varepsilon_{2t} \\ \varepsilon_{1t} \end{pmatrix}. \tag{3.37}$$

We follow the same structure for the white noise as before except that now we allow the series to be correlated to each other. Thus, the expectation of the error terms is

$$E(\varepsilon_t) = 0 \tag{3.38}$$

and the variance–covariance matrix is

$$E(\varepsilon_t \varepsilon_t') = \begin{pmatrix} \sigma_{\varepsilon_1}^2 & \sigma_{\varepsilon_1 \varepsilon_2} & \sigma_{\varepsilon_1 \varepsilon_3} \\ \sigma_{\varepsilon_2 \varepsilon_1} & \sigma_{\varepsilon_2}^2 & \sigma_{\varepsilon_2 \varepsilon_3} \\ \sigma_{\varepsilon_3 \varepsilon_1} & \sigma_{\varepsilon_2 \varepsilon_1} & \sigma_{\varepsilon_3}^2 \end{pmatrix}. \tag{3.39}$$

Note that we are allowing them to be correlated with each other as the off-diagonal terms are not necessarily zero. If we assume that there is no correlation between the error terms, then we will have

$$E(\varepsilon_t \varepsilon_t') = \begin{pmatrix} \sigma_{\varepsilon_1}^2 & 0 & 0 \\ 0 & \sigma_{\varepsilon_2}^2 & 0 \\ 0 & 0 & \sigma_{\varepsilon_3}^2 \end{pmatrix}. \tag{3.40}$$

Lack of time-lagged correlation implies

$$E(\varepsilon_t \varepsilon_{t-j}') = 0 \quad \text{for } j = \pm 1, \pm 2, \ldots, \tag{3.41}$$

where the expression on the right-hand side represents a matrix of zeros.

In matrix form, the relationship between variables x_{1t} and x_{2t} can be expressed as

$$\begin{pmatrix} x_{1t} \\ x_{2t} \\ x_{3t} \end{pmatrix} = \begin{pmatrix} \alpha_{11} & \alpha_{12} & \alpha_{13} \\ \alpha_{21} & \alpha_{22} & \alpha_{23} \\ \alpha_{31} & \alpha_{32} & \alpha_{33} \end{pmatrix} \begin{pmatrix} x_{1,t-1} \\ x_{2,t-1} \\ x_{3,t-1} \end{pmatrix} + \begin{pmatrix} \varepsilon_{1t} \\ \varepsilon_{2t} \\ \varepsilon_{3t} \end{pmatrix}. \tag{3.42}$$

In short, we can write the same model as

$$x_t \big|_{3 \times 1} = \alpha \big|_{3 \times 3} . x_{t-1} \big|_{3 \times 1} + \varepsilon_t \big|_{3 \times 1}, \tag{3.43}$$

which is exactly the same as Equation 3.20 except that x now represents a vector and α represents the matrix of coefficients (the dimensions are mentioned after the | sign for each vector and matrix).

This formulation is known as a VAR model of order 1 or simply VAR(1). There is no reason to restrict ourselves to only one lag and only three interacting variables. In general, we can consider a vector of n variables ($x_t = [x_{1t} \, x_{2t} \ldots x_{nt}]'$) with an $n \times n$ coefficient matrix $[\alpha_{ij}]$ and a vector of error terms ($\varepsilon_t = [\varepsilon_{1t} \, \varepsilon_{2t} \ldots \varepsilon_{nt}]'$). An n variable VAR(1) model can be written in the same form

$$x_t \big|_{n \times 1} = \alpha \big|_{n \times n} . x_{t-1} \big|_{n \times 1} + \varepsilon_t \big|_{n \times 1}. \tag{3.44}$$

Mentioning the dimensions for every vector and matrix would be cumbersome. An easier way to express the same model would be to consider the vector representation for an n-variable VAR(1) model:

$$\mathbf{x}_t = \alpha \mathbf{x}_{t-1} + \varepsilon_t \tag{3.45}$$

where both **x** and ε represent $n \times 1$ vectors. We can generalize this further to consider a VAR(p) model, that is, a VAR model with p lags, as follows:

$$\mathbf{x}_t = \sum_{l=1}^{p} \alpha_l \mathbf{x}_{t-l} + \varepsilon_t \qquad (3.46)$$

where α_l represents $n \times n$ matrix of coefficients for l varying between 1 and p.

The VAR literature is very well developed in economics. Many useful methods around forecasting and impulse response functions can be easily developed in this framework, which can shed light on the dynamical properties of systems. Interested readers can consult Kilian and Lütkepohl (2017).

3.2.5 Finding Moments

In this section, we want to develop some techniques for finding unconditional moments of a given ARMA process. To fix ideas, let us first consider an AR(1) process (as in Equation 3.20) with $|\alpha| < 1$. By repeated substitutions, we can express it as an MA(∞) process (Equation 3.34):

$$x_t = \sum_{j=0}^{\infty} \alpha^j \varepsilon_{t-j}. \qquad (3.47)$$

There are two useful methods for finding the moments. Let us first describe the straightforward method with the help of a simple example. Say we need to find the first moment of the variable x. We can directly take the expectation of Equation 3.20 to get

$$E(x_t) = \sum_{j=0}^{\infty} \alpha^j E(\varepsilon_{t-j}). \qquad (3.48)$$

Since ε is a white noise term and by assumption has zero mean, we can easily derive

$$E(x_t) = 0. \qquad (3.49)$$

The second method uses a trick based on the idea of stationarity. Recall that for a stationary process, the first and the second moments are constant. Therefore, by taking the expectation on the AR(1) process (Equation 3.20), we can write

$$E(x_t) = \alpha E(x_{t-1}) + E(\varepsilon_t), \qquad (3.50)$$

which can be rewritten as

$$(1 - \alpha)E(x_t) = E(\varepsilon_t), \qquad (3.51)$$

implying that

$$E(x_t) = 0. \tag{3.52}$$

Notice that we have equated the expectation terms $E(x_t) = E(x_{t-1})$, which is implied by weak stationarity.

This trick becomes more useful for finding the second moment. In the same Equation 3.20, we can apply a variance operator on both sides to derive

$$var(x_t) = \alpha^2 var(x_{t-1}) + \sigma_\varepsilon^2. \tag{3.53}$$

The covariance term between x_{t-1} and ε_t is zero since ε_t is a white noise term. By using the same trick, we see that $var(x_t) = var(x_{t-1})$ due to stationarity. Therefore, we can write

$$var(x_t) = \frac{\sigma_\varepsilon^2}{1 - \alpha^2}. \tag{3.54}$$

Analogous calculations will work for a VAR(p) process, except that one needs to account for the number of variables properly and the solution would be in matrix form.

Autocorrelation Function of the ARMA Process

In this section, we will discuss how to find autocorrelation functions (ACFs of the ARMA processes. This is a useful tool to get a sense of the underlying data-generating process. Note that the AR and the MA representations are clearly not unique as we can convert one into the other (as long as the corresponding lag poly-nomials are invertible). Thus it would be useful to get a unique representation of the process. The ACF serves two other important purposes. First, it will help us to guess the structure of an ARMA process to fit a given set of data. Note that so far we have worked with an ARMA(p, q) process by relying on the Wold decom-position theorem that an ARMA process will allow us to capture any given weakly stationary stochastic process. But we have not discussed yet how we can, given a dataset, find the best fit of an ARMA model. We will see that the ACF sheds some light on that. The second important purpose of an ACF is that, it tells us about the persistence of a process. A highly persistent process will have high autocorrelation coefficients and a process with very low persistence will have low autocorrelation coefficients.

Let us start with the simplest example of a white noise term $\varepsilon_t \sim iid(0, \sigma_\varepsilon^2)$. We can immediately see that $\gamma_0 = \sigma_\varepsilon^2$ and $\rho_0 = 1$. For all lags j larger than 0, clearly $\gamma_j = 0$ and therefore $\rho_0 = 0$.

A useful and non-trivial exercise would be to find the ACF of an AR(1) process (Equation 3.20). First, we have to find out the autocovariances at different lags. Lag-0 is the easiest:

$$\gamma_0 = var(x_t) = \frac{\sigma_\varepsilon^2}{1 - \alpha}. \tag{3.55}$$

Since the process has zero mean,

$$\gamma_1 = E(x_t x_{t-1}) = \frac{(\alpha \sigma_\varepsilon^2)}{(1-\alpha)} = \alpha \gamma_0 \qquad (3.56)$$

Continuing in the same way, we get

$$\gamma_2 = E((\alpha x_{t-1} + \varepsilon_t)x_{t-2}) = \alpha^2 E(x_{t-2}^2) = \alpha^2 \gamma_0 \qquad (3.57)$$

and so on. The pattern is obvious. We can easily derive the autocorrelation function by dividing each autocovariance value by the variance, and we can generate the ACF as follows: $\rho_1 = \alpha$, $\rho_2 = \alpha^2$, and so forth. Generally, we write

$$\rho_j = \alpha^j \qquad (3.58)$$

for all j.

There is an alternative method to generate the ACF by utilizing the MA(∞) representation. Recall that Equation 3.20 can also be written as Equation 3.32. One can try to directly find the ACF by direct substitution of the expression in the equations for autocovariances. To help with thinking about the idea, below we provide the calculations for the ACF of an MA(1) process. Consider the process given in Equation 3.22. At lag $j = 0$, we can easily find

$$\gamma_0 = \text{var}(\beta \varepsilon_{t-1} + \varepsilon_t) = (\beta^2 + 1)\sigma_\varepsilon^2. \qquad (3.59)$$

For the second equality, we have utilized the fact that ε_ts are serially uncorrelated. Next, we calculate the autocovariance at lag $j = 1$:

$$\gamma_1 = E[(\beta \varepsilon_{t-1} + \varepsilon_t)(\beta \varepsilon_{t-2} + \varepsilon_{t-1})] = \beta \sigma_\varepsilon^2. \qquad (3.60)$$

Next, we calculate the autocovariance at lag $j = 2$:

$$\gamma_2 = E[(\beta \varepsilon_{t-1} + \varepsilon_t)(\beta \varepsilon_{t-3} + \varepsilon_{t-2})] = 0. \qquad (3.61)$$

Clearly, the autocorrelation is exactly zero. Intuition is important here to understand the nature of the process. Recall that in the MA(1) process in Equation 3.22, the right-hand side consists of only two white noise terms, ε_t and ε_{t-1}. If we lag the variable x_t by two periods, then the corresponding terms would be ε_{t-2} and ε_{t-3}. Note that neither of these terms has any overlap with ε_t or ε_{t-1}. Therefore, the cross-correlation values have to be zero. Naturally, the same logic can be applied for all lags $j > 2$ as well, and it can be easily shown that

$$\gamma_j = 0 \quad \forall j > 1. \qquad (3.62)$$

Therefore, we can now write down the autocorrelation function as

$$\rho_0 = 1 \qquad (3.63)$$

$$\rho_1 = \frac{\beta}{1 + \beta^2} \qquad (3.64)$$

$$\rho_j = 0 \quad \forall j > 1. \qquad (3.65)$$

There are more general and useful techniques to find an ACF: for example, Yule–Walker equations. Here we will not describe the procedures. Interested readers may consult textbooks such as Brockwell and Davis (2016) and Hamilton (1994). Most of the software and time series toolboxes for programming language environments allow users to generate the empirical ACF from the data very easily. Such plots of ACF give us a preliminary understanding of the number of MA lags to incorporate when trying to fit an ARMA model. Also, we get a sense of persistence in the data. It should be noted in this context that the empirical ACF alone is often not precise enough to clearly differentiate between specifications of ARMA models. For that purpose, we will develop some new ideas in the rest of this chapter.

Before wrapping up this discussion, let us also mention the concept of the *partial autocorrelation* function (PACF). This measure $\pi_x(j)$ for $j \geq 2$ is defined as the coefficient β_j from the optimal linear prediction of x_t on the basis of the jth previous observation, accounting for all intermediate observations:

$$x_t = \beta_1 x_{t-1} + \beta_2 x_{t-2} + \cdots + \beta_t x_{t-j} + u_t. \tag{3.66}$$

The same equation can be easily modified to account for a process with non-zero mean as well.

This measure allows us to check the relationship between pairs of observations at a given lag *controlling for* intermediate observations. The idea behind this operation can be understood easily from the ACF of the baseline AR(1) process shown in Equation 3.20. Recall from Equation 3.58 that the ACF decays as α^j for the jth lag. Therefore, say if $j = 4$, we will have $\rho_4 = \alpha^4$. But given the form of the process in Equation 3.20, there is no direct relationship between the observations made at times t and $t - 4$. Partial autocorrelation captures precisely this point. One can show that for AR(p) processes, the partial autocorrelation after p lags will be zero, whereas for MA(q) processes it will continue indefinitely. In this way, it mirrors the properties of the standard ACF for which the values are zeros after q lags for MA(q) processes and continue indefinitely for AR(p) processes. It is useful to gather these properties in one place for reference. In Table 3.1, we have described the properties of the ACF and the PACF.

As an example, in Figure 3.3 we show the sample ACF and PACF of the GDP growth rate shown in Figure 3.1. To keep the figure legible, we have not plotted the error bars for the point estimates. Many of the point estimates may turn out to be statistically insignificant, especially at higher lags.

3.2.6 Estimation Procedure

So far we have discussed the ARMA and its multivariate generalization as theoretical objects and analyzed their properties. Now we are in a position to ask the most important question from a data-oriented point of view: Given a series of observations $X_T = \{x_1, x_2, \ldots, x_T\}$, how can we fit an ARMA(p, q) model? Also, if

Table 3.1 *Description of the visually identifiable properties of the autocorrelation functions and the partial autocorrelation functions for stationary AR(p), MA(q), and ARMA(p, q) models*

Process	ACF ($\rho(j)$)	PACF ($\pi(j)$)
AR(p)	Infinite length; dampened exponential or sinusoidal curve	Finite length; for lags $j > p$, $\pi(j) = 0$
MA(q)	Finite length; for lags $j > q$, $\rho(j) = 0$	Infinite length; dampened exponential or sinusoidal curve
ARMA(p, q)	Mimics AR(p) for lags $j > q$	Mimics MA(q) for lags $j > p$

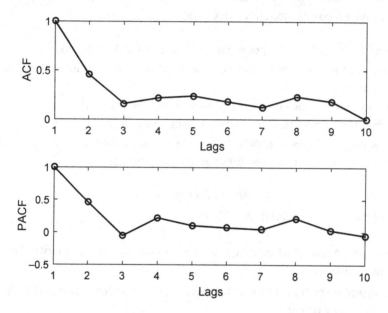

Figure 3.3 Sample ACF and PACF of the GDP growth rate shown in Figure 3.1.

we have a set of multivariate observations, how can we fit a VAR model to it? A multivariate ARMA model can also be estimated in the exact same way. However, it is more commonplace to see VAR models rather than VARMA models since a VARMA model requires a much larger number of parameters to be estimated and often the lagged dependence on the error terms do not have theoretical justifications. In the present context, we will focus only on ARMA models and VAR models, as is standard in the literature.

Let us start from the univariate context. We rely on the fundamental result obtained from the Wold decomposition theorem that any weakly stationary data-generating process can be approximated by an ARMA(p, q) process. There are three main steps in the estimation procedure. First, we need to determine the order (p and q) of the process. Second, we need to find the values of the coefficients and estimate the residuals. Third, we then perform tests to check the validity of the estimated model. This procedure goes back to an original formulation by Box and Jenkins in the 1970s (a recent textbook exposition is given in Box et al. [2015]) and has evolved into a standard method.

Box–Jenkins Methodology

The Box–Jenkins method is fundamentally concerned with model estimation as well as model selection. The basic recursive model-selection technique works as follows. Given a covariance stationary time series:

   (i) Identify the orders of a potentially suitable ARMA(p, q) model.
  (ii) Estimate the corresponding coefficients under the assumption that the orders are accurate.
 (iii) Perform diagnostics on the fitted model and the residuals.
 (iv) If the fitted model fails the tests, go back to the first step.
  (v) If the model seems satisfactory in terms of the statistical tests, one can use the model for further analysis: for example, for forecasting.

Model Identification

There are the steps for model identification.

   (i) First, given the set of data $\{x_t\}$, we need to check for stationarity. This can be done in two ways:
       1. Visual inspection of the data often tells us immediately whether the data is stationary or not.
       2. One can also utilize statistical tests to check for stationarity. We will discuss some well-known statistical tests, such as the Dickey–Fuller test later in this chapter.
  (ii) If the data series is non-stationary, transform the data to achieve stationarity (see Section 3.3 for a discussion on non-stationarity).
       1. For example, GDP per capita (G_t) can be non-stationary but the growth rate ($g_t = log(G_t) - log(G_{t-1})$) can be stationary.
       2. Sometimes, more than one round of differencing might be required to achieve stationarity. Price indices (such as the consumer price index) often display this kind of behavior.

(iii) The next step is to select the orders p and q.

1. First, compute the empirical ACF and PACF. Most standard software packages generate empirical ACF and PACF very easily.
2. These plots will give an idea of the orders of the MA and AR lags, respectively.
3. However, the empirical ACF and PACF are often non-informative. One can utilize more sophisticated statistical criteria as well (such as AIC and BIC, described below).

(iv) Once we have fixed the orders p and q, we know the number of parameters to be estimated.

1. For AR(p) processes, ordinary least square estimation can be applied.
2. For general ARMA(p, q) processes, maximum likelihood estimation can be applied.

(v) Next, we check for autocorrelation in the residuals. The idea is that the residuals of a good model should not be autocorrelated. All autocorrelated components in the data should already be explained by the model itself.

1. One can perform the Ljung–Box test. Other possibilities are the Breusch–Godfrey or Durbin–Watson tests.
2. If the residuals do not have any autocorrelation, then the model is acceptable.
3. Otherwise, we need to modify the specification by going back to the first step above.

Estimation of the Lag Orders

Note that the above procedure depends crucially on the idea that we can recursively define the models and check for their suitabilities. Thus if we started with a small number of lags, but we find that the original model is not suitable, we may want to go for a model with larger number of lags, and vice versa. There are two factors that we have to be careful about while choosing the lags. At one extreme, we might pick very large p and q leading to overfitting the data. In such cases, we would have high in-sample fit but poor out-of-sample performance. At the other extreme, we might pick very small p and q. Then the model would underfit the data leading to low in-sample fit but better out-of-sample performances. Either way it is bad since the maximum likelihood estimator would not be consistent with a mis-specified ARMA(p, q) model. We have already noted that using ACF and PACF gives us a preliminary understanding of lag orders. But visual inspection may not be very accurate, especially for small samples, and visual inspection is certainly not a good substitute for a formal statistical test. In the next subsection, we will discuss the usage of information criteria to resolve this issue.

Information Criteria

Information criteria are a commonly used tool for model selection from a set of competing models that can explain the same set of data to varying degrees. The basic goal of this exercise is to minimize some kind of *information loss* to come up with a better model. A model which loses lesser amount of information can be taken to be a better model (see also the discussion in Section 2.3.7). If we over-parameterize, then the in-sample fit has to necessarily improve, but that is not good from the perspective of using the model for prediction. Intuitively, if we try to fit each and every squiggle on a curve, then the fitted model will perform very badly for capturing the average behavior of the process in out-of-sample calculations.

Let us denote the fit of the model by the variance of the residuals ($\hat{\sigma}_{p,q}^2$). Clearly, if the variance of the residuals reduce, then the model has better in-sample fit. Thus there would be a tendency to increase p and q to reduce the variance. The objective is to penalize higher values of p and q and correct for $\hat{\sigma}_{p,q}^2$ to attain a balance. The problem statement boils down to minimizing information criteria IC with respect to p and q where the information criteria are defined as the variance of the residuals along with a penalty term for over-parametrization (Neusser, 2016, pp. 101):

$$\text{Akaike IC (AIC)} = \log(\hat{\sigma}_{p,q}^2) + \frac{2}{T}\left(p+q\right), \tag{3.67}$$

$$\text{Bayesian or Schwarz IC (BIC)} = \log(\hat{\sigma}_{p,q}^2) + \frac{\log(T)}{T}\left(p+q\right), \tag{3.68}$$

and

$$\text{Hannan–Quinn IC (HQIC)} = \log(\hat{\sigma}_{p,q}^2) + \frac{2(\log(\log(T)))}{T}\left(p+q\right). \tag{3.69}$$

While there are many other model-selection techniques, these three measures are standard and most software packages would produce these values. It might be noted that, as a general rule, BIC and HQIC penalize over-parameterization more than AIC and hence typically economize on the number of parameters. However, simulation results indicate that AIC might pick a better model (closer to the true data-generating process) in small samples. Finally, we note that none of these criteria clearly dominates the rest.

In Table 3.2, we provide the estimated AIC and BIC values from ARMA(p, q) models on the GDP growth rate data from Figure 3.1. Each cell refers to the (AIC,BIC) value of the corresponding ARMA(p, q) model: that is, the first cell can be read as ARMA(0,0) model having an AIC value of 225.05 and a BIC value of 229.2. In each cell, the row value corresponds to lag p (AR(p)) and the column value corresponds to lag q (MA(q)). Both lags vary from 0 to 3. One can create a larger model as well. While a larger model fits the data better, most of the

Table 3.2 *AIC and BIC for ARMA(p, q) estimation on the GDP growth rate data from Figure 3.1*

ARMA(p, q)	0	1	2	3
0	(225.05,229.20)	(213.03,219.26)	(214.81,223.12)	(216.35,226.74)
1	(213.32,219.56)	(214.68,222.99)	(213.73,224.12)	(214.93,227.40)
2	(215.15,223.46)	(216.21,226.59)	(215.22,227.69)	(216.17,230.72)
3	(214.31,224.69)	(214.71,227.17)	(216.67,231.21)	(218.02,234.64)

time it does not permit any interpretation. Generally, the preference is for choosing smaller models. In terms of AIC and BIC, AR(1) and MA(1) models seem to be good contenders for the best model in the present context. However, in many cases, the model choice due to AIC and BIC do not coincide. An additional criterion could be model interpretation. If a model is more economical than others in terms of parameters or more easily explainable, and if it has comparable AIC/BIC values, then one can go for that model.

Estimation of ARMA(p, q) Parameters

So far we have described the procedure of lag selection. Now we will describe the final step of estimating the parameters once the lags have been optimally selected. We need to provide a clarification here. Computer programs often do the estimation of parameters jointly with information criteria among many other variables. In Table 3.2, for example, we have provided only the AIC and BIC. But for each estimation we already had the estimated parameter values as well. Here it is only for descriptive purposes that we are describing parameter estimation sequentially after describing information criteria.

We will proceed in two steps. First, we will discuss two methods to estimate AR(p) processes, and then we will discuss a general maximum-likelihood-based method for estimating ARMA(p, q) processes. There are two reasons for describing the first two estimation techniques. First, the methods shed light on the working of AR(p) processes in two different ways, and that is often useful to know. Second, such a discussion also shows that AR(p) processes are amenable to multiple types of estimation including the standard least square method. The following discussion has been partially inspired by the discussion of the estimation of ARMA models in Neusser (2016).

Let us say we have a model of the form (the same as in Equation 3.23)

$$x_t = \sum_{i=1}^{p} \alpha_i x_{t-i} + \varepsilon_t. \tag{3.70}$$

The goal is to estimate the parameter vector $\{\alpha_1, \alpha_2, \ldots, \alpha_p\}$ from a given set of observations X_t. We will describe two methods to estimate the AR(p) process: the Yule–Walker procedure and the ordinary least square procedure.

Yule–Walker method: The easiest way to see the working of this method is to simply multiply Equation 3.70 by x_{t-k} for $k = 0, \ldots, p$ on both sides and applying an expectation operator to them. For example, multiplying by x_t and applying expectation would give us

$$\gamma_0 = \sum_{i=1}^{p} \alpha_i \gamma_i + \sigma_\varepsilon^2. \tag{3.71}$$

Multiplying by x_{t-1} and applying expectation would give us

$$\gamma_1 = \sum_{i=1}^{p} \alpha_i \gamma_{i-1}. \tag{3.72}$$

Note that x_{t-1} has no matching term for ε_t and therefore the covariance value is zero. In this way, we can keep on multiplying and generate $p+1$ equations involving autocovariances of $\{x_t\}$ and the parameter vector $\{\alpha_1, \ldots, \alpha_p, \sigma_\varepsilon^2\}$. Solving this set of equations allows us to estimate values of the parameter vector $\{\hat{\alpha}_1, \ldots, \hat{\alpha}_p, \hat{\sigma}_\varepsilon^2\}$.

Ordinary least square (OLS) method: Consider the same process as in Equation 3.70. We can treat it directly as a regression model and estimate it via the OLS method. We can assume that x_t is the endogenous variable, $x_{t-1}, x_{t-2}, \ldots, x_{t-p}$ are exogenous variables, ε_t is the error term, and finally the coefficient vector $\{\alpha_1, \ldots, \alpha_p\}$ is unknown and to be estimated. To apply the OLS method, we augment the model with a constant α_0.

$$x_t = \alpha_0 + \sum_{i=1}^{p} \alpha_i x_{t-i} + \varepsilon_t \tag{3.73}$$

Let us collect terms in the form of a matrix. Let

$$Y = \begin{bmatrix} x_{p+1} \\ \vdots \\ x_T \end{bmatrix}, \quad \beta = \begin{bmatrix} \alpha_0 \\ \vdots \\ \alpha_p \end{bmatrix}, \quad \varepsilon = \begin{bmatrix} \varepsilon_{p+1} \\ \vdots \\ \varepsilon_T \end{bmatrix}, \tag{3.74}$$

and

$$X = \begin{bmatrix} 1 & x_p & x_{p-1} & x_{p-2} & \cdots & x_1 \\ \vdots & \vdots & \vdots & \vdots & \cdots & \vdots \\ 1 & x_{T-1} & x_{T-2} & x_{T-3} & \cdots & x_{T-p} \end{bmatrix}. \tag{3.75}$$

Combining these matrices, we can express Equation 3.73 in a more convenient form:

$$Y = X\beta + \varepsilon. \tag{3.76}$$

Therefore, the OLS estimator is

$$\hat{\beta} = (X'X)^{-1}X'Y. \tag{3.77}$$

The derivation of the OLS estimator follows from Section 2.3.5 in Chapter 2. Interested readers may additionally consult Hamilton (1994) for a time series perspective and Greene (2003) for an econometric exposition.

There are two technical issues with this estimator. First, the regressors (i.e. the variables on the right-hand side of Equation 3.73) are clearly correlated with error terms. Second, the first p observations $(x_1, x_2, \ldots, x_p)$ affect the OLS estimates. However, as Theorem 5.2 in Neusser (2016) (see also Brockwell et al., 1991) shows, $\hat{\beta}_{OLS}$ is asymptotically normal and asymptotically equivalent to the Yule–Walker estimator (Neusser, 2016, p. 92). For ARMA(p, q) models, OLS estimation is problematic since the error terms $(\varepsilon_{t-1}, \ldots, \varepsilon_{t-q})$ are not observable.

Maximum likelihood estimator: To estimate a general ARMA(p, q) model, we utilize maximum likelihood estimation. Let us say we have a model of the form (the same as in Equation 3.25 with $\alpha_0 = 0$)

$$x_t = \sum_{i=1}^{p} \alpha_i x_{t-i} + \sum_{j=1}^{q} \beta_j \varepsilon_{t-j} + \varepsilon_t \tag{3.78}$$

with normally distributed white noise ε_t, that is, $\varepsilon_t \sim N(0, \sigma^2)$. Note that the maximum likelihood method is parametric in nature and we have to start from some distributional assumptions. The most standard assumption is normality. Following Neusser (2016, pp. 95), we can write the exact Gaussian likelihood function as

$$L(\beta|x) = (2\pi)^{-T/2}|\Gamma(\beta)|^{-1/2}exp\left(-\frac{1}{2}x'\Gamma(\beta)^{-1}x\right) \tag{3.79}$$

where $\Gamma(\beta) = E(xx')$ is the $T \times T$ covariance matrix of x which is a function of β. Theorem 5.3 from Neusser (2016) shows that the maximum likelihood estimator is asymptotically normally distributed. Finally, this estimator is also asymptotically efficient.

The actual computation of the maximum likelihood estimator can be quite involved and we have to resort to numerical techniques to maximize the likelihood function. All popular software/programming languages with time series packages (R, MatLab, Python, Stata, etc.) have these kinds of programs built in.

Financial Networks from Multivariate Time Series

Now that we have developed the concept of the estimation of stationary processes, here we will discuss an application of the concept: constructing a financial network from a multivariate return series. We will discuss the concept of networks and the associated statistics in Chapter 5. Here, we will treat a network as simply a

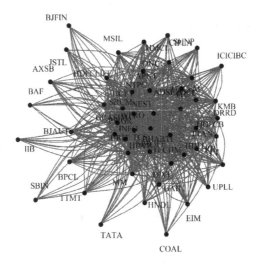

Figure 3.4 Stock return correlation network constructed from intra-day data of Indian firms on a given day (National Stock Exchange of India, December 2020).

collection of a set of nodes and the linkages between them. Consider a stock market where N number of stocks are being traded. Each of them will have a price at any given point in time t. We assume that t is discrete and goes from 1 to T. Let us denote the price of the ith asset at time point t as p_{it} – say this denotes the daily closing price of that stock. This price series itself will be non-stationary. Let us consider the log-return series arising out of it:

$$r_{it} = \log(p_{it}) - \log(p_{i,t-1}) \tag{3.80}$$

for all $i = 1, 2, \ldots, N$ and $t = 1, 2, \ldots, T$. This kind of operation of differencing is discussed in more detail in Section 3.3.

Cross-Correlation Network

A standard way to construct a network out of a multivariate time series is to consider each stock to represent one node and their correlations (actually a transformation of that as we will discuss later in this section) to represent the linkages. From N stocks, we can construct a correlation matrix ρ of size $N \times N$. The correlation value itself is not a metric, as it can be negative. A standard method is to convert it into a metric by taking a simple transformation

$$w_{ij} = \sqrt{2(1 - \rho_{ij})}. \tag{3.81}$$

This metric was proposed by Gower (1966) (see Mantegna and Stanley, 2007).

In Figure 3.4, we show one example of a financial network constructed from intra-day data, sampled at 10-s intervals, of the largest 50 stocks traded in the National Stock Exchange of India which constitute the NIFTY 50 index, sampled from the December 2020 listing. Since the full network with N nodes, denoting N

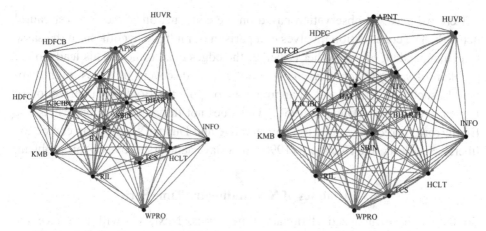

Figure 3.5 Granger causal networks of 15 Indian companies constructed from their intra-day stock returns. The left panel shows the network evaluated at 5% level of significance, whereas the right panel shows it evaluated at 10% level of significance.

number of stocks, will have a large number of edges linking the nodes – specifically, $\binom{N}{2}$ number of edges where $N = 50$ – we have applied a threshold for the purpose of visualization.

Lagged Comovement Network

Here we introduce the concept of lagged comovement and Granger causality. Consider a bivariate analogue of the vector autoregression model we looked at in Section 3.2.4 –

$$\begin{pmatrix} x_{1t} \\ x_{2t} \end{pmatrix} = \begin{pmatrix} \alpha_{11} & \alpha_{12} \\ \alpha_{21} & \alpha_{22} \end{pmatrix} \begin{pmatrix} x_{1,t-1} \\ x_{2,t-1} \end{pmatrix} + \begin{pmatrix} \varepsilon_{1t} \\ \varepsilon_{2t} \end{pmatrix}. \tag{3.82}$$

If α_{12} is significant, then we say that x_2 Granger causes (systematically predicts) x_1. If α_{21} is significant, then the direction of causality is reversed. It is possible that neither of the two series causes the other or that both do. In general, one can allow for a larger number of lags and variables.

This concept allows us to construct a lagged correlation network in the form of a Granger causal network. Here we show an example. We take 15 randomly chosen companies featuring in the stock network in Figure 3.4 and construct a Granger causal network out of them. We note two things here. First is that, in general, such a network will be directed: links will have directions from one node to the other based on whether or not one stock has a Granger causal effect on the other. Second, the edges are binary. They either exist or not based on whether the relationships are significant or not. In Figure 3.5 we show the Granger causal networks at 5% and 10% levels of significance.

Here we make an observation based on the construction of the Granger causal network. The construction involves comparison of multiple hypothesis at the same time independent of each other (whether the edges exist or not). This leads to the problem of *multiple comparisons* where the false discovery of edges might happen. In such cases, the classical approach is to apply the Bonferroni correction or Duncan's correction (Duncan, 1955). This does not fully allay the concern, as the Bonferroni correction can be too conservative and Duncan's correction can be too liberal. See Berry and Hochberg (1999) for a Bayesian perspective on this problem.

3.3 Analyses of Non-stationary Time Series

So far we have analyzed stationary time series. Next, we will introduce the concept of non-stationary time series. We can express the reasons for studying non-stationary time series in multiple ways. But the best way is to simply refer to the famous statement by D. J. Thomson, who said: "Experience with real-world data, however, soon convinces one that both stationarity and Gaussianity are fairy tales invented for the amusement of undergraduates" (Thomson, 1994)!

Non-stationarity might arise in a number of ways. The time series can have a deterministic trend (exhibiting trend stationarity), there can be regime shifts in level, or there can be changes in the variance. Finally, the time series can possess unit roots (exhibiting a stochastic trend). In the present discussion, we will examine time series with deterministic and stochastic types of non-stationarity. For the sake of discussion, we will ignore seasonal fluctuations in the following examples. For example, both the GDP series shown in Figure 3.1 and the evolution of the market capitalization of Bitcoin shown in Figure 3.2 are non-stationary. We will describe the exact meaning of this term below. A process $\{x_t\}$ is called trend stationary if

$$x_t = f(t) + \zeta_t \tag{3.83}$$

where $f(.)$ is a function of t and ζ_t is a stationary process. For example, let us say ζ_t is some stationary ARMA(p, q) process and x_t has the form

$$x_t = \sum_{i=0}^{k} a_i t^k + \zeta_t \tag{3.84}$$

for some positive integer k and constant terms a_i. Then x_t is a trend stationary process. On the other hand, a process x_t is called difference stationary if

$$\Delta^d x_t = \zeta_t \tag{3.85}$$

where Δ denotes a difference operator such that $\Delta \omega_t = \omega_t - \omega_{t-1}$ for a process ω_t, d is the order of differencing, and ζ_t is stationary. This process is very useful for many reasons. The best-known case is probably that of the random walk. Let's say, $d = 1$ and ε_t is an IID variable. Note that an IID variable is stationary by definition and therefore ε_t is a candidate for ζ_t in Equation 3.85 above. Then clearly the above equation reduces to an AR process with the AR coefficient being 1:

$$x_t = x_{t-1} + \varepsilon_t, \tag{3.86}$$

which is a simple random walk process. We will discuss below how difference stationarity can be modeled.

3.3.1 AR Process with Unit Root

Here we will analyze the process from the point of view of the ARMA framework we have developed before. Consider the process described in Equation 3.86. Note that the corresponding lag polynomial can be written as

$$\alpha(\mathcal{B}) = 1 - \alpha_1 \mathcal{B} \tag{3.87}$$

where $\alpha_1 = 1$. The corresponding characteristic polynomial is $\alpha(z) = 1 - z$. Evidently, it has a unit root since $\alpha(1) = 0$. By backward substitution, we get (assuming that the process started finite periods back at time point $t = 0$)

$$x_t = x_0 + \varepsilon_1 + \cdots + \varepsilon_t. \tag{3.88}$$

An important point to notice here is that the effect of any shock remains forever. The best way to see this is to consider an AR(1) process with a coefficient less than 1 (as in Equation 3.20):

$$x_t = \alpha^t x_0 + \alpha^{t-1}\varepsilon_1 + \cdots + \varepsilon_t. \tag{3.89}$$

If $|\alpha| < 1$, then the effect of any shock on x_t through ε_k for some k clearly decays over time as the multiplier α^{t-k} goes to zero. Since for a random walk we have $\alpha = 1$, the effect of the shock does not decay and is described as "permanent."

From the above expression (Equation 3.88), we can easily find the variance of the process:

$$Var(x_t|x_0) = \sum_{j=1}^{t} \sigma_\varepsilon^2 = t\sigma_\varepsilon^2. \tag{3.90}$$

The autocovariance is given by (assuming $t > s$ and x_0 is known)

$$Cov(x_t, x_s|x_0) = E((x_t - x_0)(x_s - x_0)|x_0) = (t - s)\sigma_\varepsilon^2. \tag{3.91}$$

Combining the above two expressions, we can find the autocorrelation as

$$\rho(x_t, x_s|x_0) = \frac{Cov(x_t, x_s|x_0)}{\sqrt{V(x_t|x_0)V(x_s|x_0)}}$$

$$= \sqrt{\frac{t-s}{t}} \tag{3.92}$$

These moments are all time dependent, making the process non-stationary. In particular, variance increases linearly with time. Often the random walk process is augmented with a drift term δ:

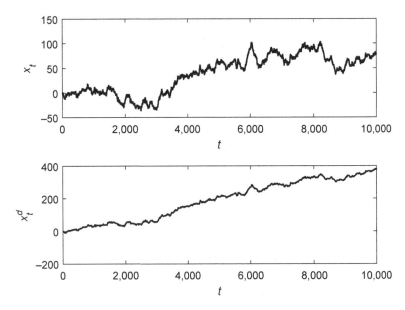

Figure 3.6 Simulation of random walk (top panel). The bottom panel shows the same time series added with a small drift in the upward direction.

$$x_t = \delta + x_{t-1} + \varepsilon_t. \tag{3.93}$$

By backward substitution, we can write the expanded process as

$$x_t = x_0 + \delta t + \sum_{j=1}^{t} \varepsilon_j, \tag{3.94}$$

which is an augmented version of Equation 3.88.

We show an example of a random walk with Gaussian noise in Figure 3.6. The divergence of the process is clearly visible. The bottom panel exhibits the same series with a small upward drift. These types of process visually mimic stock price data very well, although they do not reproduce a number of specific regularities found in the stock markets. We will discuss some of them in the next section.

3.3.2 Unit Root Testing

Here we describe a few statistical tests that allow us to check for stationarity of a process (or the absence of non-stationarity). Fundamentally, the objective is to estimate an autoregressive model on a given set of data and test whether there is a unit root or not. While this task is seemingly trivial, non-triviality arises from the fact that the asymptotic distribution of the estimate for the unit root does not converge to normal distribution and therefore we cannot perform the inference exercise with standard t-tests (as in Chapter 2). This problem was recognized long ago, and currently there is a wide array of tests. We will not get into a complete enumeration of all such tests. Instead we will describe the family of Dickey–Fuller tests that is probably the most standard solution to this problem. For more elaborate discussions and relevant econometric background, readers may consult Hamilton (1994), Brockwell et al. (1991), Neusser (2016), and Tsay (2010).

Dickey–Fuller (DF) Test: Consider an AR(1) model (Dickey and Fuller, 1979):

$$x_t = \alpha x_{t-1} + \varepsilon_t. \tag{3.95}$$

A straightforward hypothesis test can be conducted on

$$H_0: \quad \alpha = 1 \quad \text{against} \quad H_1: \quad \alpha \in (-1, 1). \tag{3.96}$$

This is a one-sided test. The DF test statistic is simply the t-ratio of the estimated α:

$$\hat{t} = \frac{\hat{\alpha} - 1}{se(\hat{\alpha})} \tag{3.97}$$

where the denominator is the standard error of the estimated coefficient. This test differs from the usual hypothesis testing on parameter estimation in that the Dicky–Fuller distribution does not converge to a normal distribution and this needs to be simulated. Standard software packages apply simulation to the Dicky–Fuller distribution automatically and provide the resulting significance.

Note that the process given by Equation 3.95 can also be written as

$$\Delta x_t = \xi x_{t-1} + \varepsilon_t \tag{3.98}$$

where $\xi = \alpha - 1$. Therefore, the hypothesis test can also be conducted on

$$H_0: \quad \xi = 0 \quad \text{against} \quad H_1: \quad \xi \in (-2, 0). \tag{3.99}$$

The DF test statistic in this case can be simplified to the following expression:

$$\hat{t} = \frac{\hat{\xi}}{se(\hat{\xi})}. \tag{3.100}$$

Augmented Dickey–Fuller test: Suppose we have an AR(p) process:

$$x_t = \alpha_1 x_{t-1} + \alpha_2 x_{t-2} + \cdots + \alpha_p x_{t-p} + \varepsilon_t. \tag{3.101}$$

This can be rewritten in terms of first differences as

$$\Delta x_t = \xi x_{t-1} + \theta_1 \Delta x_{t-1} + \cdots + \theta_{p-1} \Delta x_{t-p+1} + \varepsilon_t \tag{3.102}$$

where ξ, θ_1, θ_2, $\ldots, \theta_{p-1}$ are defined suitably. The hypothesis testing can be conducted on

$$H_0: \quad \xi = 0 \quad \text{against} \quad H_1: \quad \xi \in (-2, 0). \tag{3.103}$$

The ADF test statistic is

$$\hat{t} = \frac{\hat{\xi}}{se(\hat{\xi})}. \tag{3.104}$$

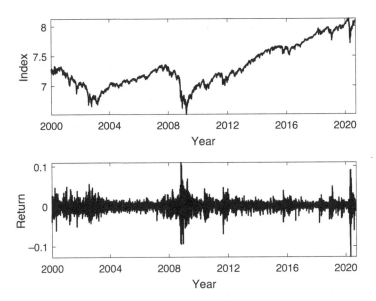

Figure 3.7 Stock index fluctuations. Upper panel: Evolution of log of the S&P 500 index series from January 3, 2000 to July 28, 2020. Lower panel: The corresponding return series in the form of first difference.

We can easily augment the process with a constant term c:

$$\Delta x_t = c + \xi x_{t-1} + \theta_1 \Delta x_{t-1} + \cdots + \theta_{p-1} \Delta x_{t-p+1} + \varepsilon_t, \tag{3.105}$$

or a trend term:

$$\Delta x_t = c + \kappa t + \xi x_{t-1} + \theta_1 \Delta x_{t-1} + \cdots + \theta_{p-1} \Delta x_{t-p+1} + \varepsilon_t. \tag{3.106}$$

Standard time series software packages allow this kind of hypothesis testing quite easily. As opposed to the Dickey–Fuller test (also the Phillips–Perron test – see Phillips and Perron [1988]), test which takes the presence of unit roots to be the null hypothesis, one can also test for non-stationarity in the data using the KPSS test (short form of Kwiatkowski–Phillips–Schmidt–Shin; see Kwiatkowski et al., 1992) which considers the presence of a unit root as the alternative hypothesis. Elliott et al. (1996) developed the ADF–GLS test, which is an improvement on the ADF test. Ng and Perron (2001) developed a *modified* Phillips–Perron test, which in turn improves upon the ADF–GLS test.

3.4 Modeling Fluctuations

Financial asset return data exhibit a number of interesting patterns (Sornette, 2009). Figure 3.7 shows the evolution of the log of the Standard & Poor's (S&P) 500 index daily data from 2000 to 2020 (data obtained from Bloomberg). Looking at the data on the level (top panel) does not tell us much beyond the presence of large fluctuations and an upward trend. In the bottom panel, we have drawn the first difference of the same dataset. This new transformed time series represents the log return of the S&P 500 index. As can be seen, return data exhibits episodes of high volatility followed by periods of low volatility. Two periods that stand apart from the rest are that of 2008–2009 (the time of the global financial crisis) and the period around March 2020 (the beginning of the COVID-19 pandemic). Large fluctuations during these periods indicate high volatility in return in the financial markets.

Such features are not specific to only S&P 500 data and, in fact, can be found in stock market data from all over the world. In the literature, these are known as *stylized facts*. In a highly cited research article, Cont (2001) collected a wide range of such stylized facts from financial markets, ranging from zero autocorrelation in asset returns to properties of the second moment of asset returns. In this section, we will study two of those features: *volatility clustering* and the *slow decay of autocorrelation of modulus of return*. Volatility clustering refers to the pattern in the data showing that volatile periods tend to cluster in time. Put differently, a volatile event is more likely to be followed by another volatile event rather than a non-volatile event. In what follows, we will analyze such a scenario from a modeling point of view. In particular, we will introduce a famous class of models called GARCH models that are able to replicate this feature in financial data (along with other features, such as fat-tailed return distribution). It is worth emphasizing that this model does not explain *why* there is volatility clustering; it merely allows us to *replicate* the pattern in the data, assuming that volatility tends to cluster. After setting up the GARCH model and describing how such models are estimated, we will discuss the phenomenon of long-range correlations and, in particular, we will relate this to the observation of the slow decay of autocorrelation of the magnitude of returns.

3.4.1 The GARCH Model: Stock Returns and Volatility Clustering

The full name of the GARCH model is the *generalized autoregressive conditional heteroscedasticity* model. This model captures two features of financial data. First, asset return data often shows time-varying volatility. In fact, it is rare to find asset return data having a constant variance over time. Second, volatility itself seems to be quite persistent, implying that a volatile period is often followed by another volatile period. Notice that the ARMA model that we have developed in the earlier sections of this chapter, is not as useful tool in this kind of a scenario as the model does not allow for time varying conditional moments.

To visually examine the idea of volatility clustering, in Figure 3.8 we present the autocorrelation functions estimated from the daily closing price data in S&P 500 times series (as we used in Figure 3.7). The top panel shows the autocorrelation function of the log return, the middle panel shows the same for absolute values of

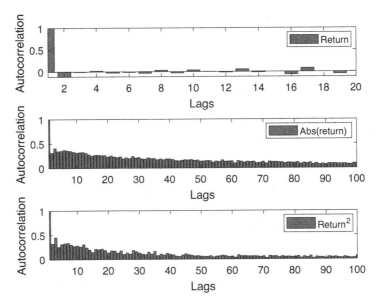

Figure 3.8 Volatility clustering in financial data. Upper panel: Sample autocorrelation function of the raw log return data from Figure 3.7. Middle and lower panels: Sample autocorrelation functions of the absolute returns and the squared returns, respectively.

the log return series, and the bottom panel shows the same for squared log return series. As is evident from the figure, the autocorrelation function drops to zero almost immediately in case of log return; however, it tends to decay very slowly for the absolute values as well as squared values of the return series. This is a unique feature of financial data that indicates persistence in volatility.

The GARCH model, which is a generalization of the ARCH model (the *autoregressive conditional heteroscedasticity* model), is a clever way to capture this statistical pattern in the data. Essentially it imposes an ARMA kind of a structure on the conditional volatility of the variable being modeled. Thus high persistence in the ARMA structure implies that the volatility series itself is persistent. We must be careful here to differentiate between conditional and unconditional volatilities. As the name suggests, GARCH models the evolution of *conditional volatility*: that is, volatility conditional on past realization of the volatility and the observed variable (Andersen et al., 2014).

Let us first look at the structure of a GARCH model to see how it works. A simple example of GARCH(1,1) is as follows:

$$r_t = \sigma_t \varepsilon_t$$
$$\sigma_t^2 = \omega + \alpha r_{t-1}^2 + \beta \sigma_{t-1}^2 \qquad (3.107)$$

where r_t is the asset return at time t (time is discrete here as was the case in the ARMA setup), ε_t is an independent standard normal random variable, and σ_t^2 is the conditional variance that evolves as a function of squared past return r_{t-1}^2 and σ_{t-1}^2. The term ω is a positive constant.

The important point to note here is that if α and β are both zero, then we are back in the world of constant variance. In order to induce conditional *heteroscedasticity* (which means different degrees of variability across observations), we need non-zero α and β. In particular, positive and high values of them would indicate high persistence in conditional volatility.

Another important feature of the model is the way it treats conditional volatility. Note that from the data we cannot directly observe volatility. For example, if we consider daily closing return data $\{r_t\}$, then for a given day t there is only one observation r_t. Therefore, obviously the corresponding variance of that single data point is zero. If we calculate day-wise in-sample volatility, then we will simply get a sequence of zeros, which is clearly not a useful way to think about how volatility evolves. A non-trivial way to use this approach would be to consider the volatility of observations in a moving window, say across a fixed number of days. The problem with such an approach is that it is too ad hoc, with no clean way to select the window length, and also potentially the resulting volatility would be non-robust with respect to the window length. The way this model considers volatility is to treat it as a *latent* variable that evolves over time but is not directly observable. Hence, given a set of data, we have to estimate its values over time.

A GARCH process with generalized number of lags can be written as (Engle and Bollerslev, 1986)

$$r_t = \sigma_t \varepsilon_t$$

$$\sigma_t^2 = w + \sum_{i=1}^{p} \alpha_i r_{t-i}^2 + \sum_{j=1}^{q} \beta_j \sigma_{t-j}^2 \tag{3.108}$$

which can be recognized as a straightforward generalization of Equation 3.107 by allowing p and q lags on the past values. It is customary to assume that $\omega > 0$, $\alpha_i \geq 0$, and $\beta_j \geq 0$ for $i = 1, \ldots, p$ and $j = 1, \ldots, q$.

We need to impose some other restrictions in order to estimate the model. As usual, ε_t is a white noise term: that is, $\{\varepsilon_i\} \sim \text{IID}(0,1)$. We also need $(\sum_i^p \alpha_i + \sum_j^q \beta_j) < 1$ for uniqueness and stationarity. Here we should clarify one more point: while GARCH models account for time-varying volatility, the resulting time series are stationary. The time-varying feature of volatility is applied only to *conditional* volatility. It can be shown that the unconditional variance of GARCH(p, q) (Equation 3.107) is

$$E(r_t^2) = \frac{\omega}{1 - \sum_i^p \alpha_i - \sum_j^q \beta_j}, \tag{3.109}$$

which would be a positive constant if the above assumptions hold. With zero mean and zero serial correlation of r_t (we will not prove these properties here; see, for

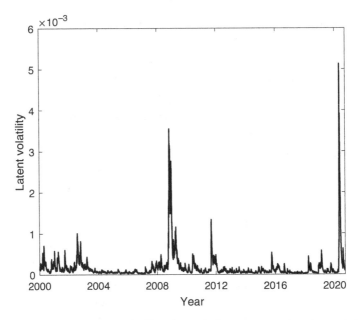

Figure 3.9 Estimated latent volatility from the log return series shown in Figure 3.7. Volatility spikes are obvious during the times of the financial crisis and the COVID-19 pandemic.

example, Tsay [2010] for a more elaborate treatment), this indicates that the process would in fact be weakly stationary. For a review of ARCH-type processes, interested readers can consult Bera and Higgins (1993).

Estimation of GARCH models: Like other models we have discussed, GARCH models can also be estimated using maximum likelihood (Tsay, 2010). Most of the standard time series software/packages will contain routines to perform the estimation exercise.

In Figure 3.9 we show the estimated latent volatility from the same log return dataset as depicted in Figure 3.7. The estimated latent volatility successfully captures the effects of the 2008–2009 crisis in the financial markets and the COVID-19 pandemic.

Latent volatility is known to influence pairwise correlations in the stock market. In Figure 3.10 we show the same network as in Figure 3.4. The only difference is that, in this case, we have normalized each return series by its corresponding volatility series estimated with a GARCH(1,1) model. The network retains its general properties even after the adjustment.

3.5 Scaling and Long Memory

The idea of scaling has been explored in great detail in the physics literature. Over time, it has found applications in widely different fields ranging from geophysics to finance. Benoit Mandelbrot was the first scientist to bring the concept of scaling

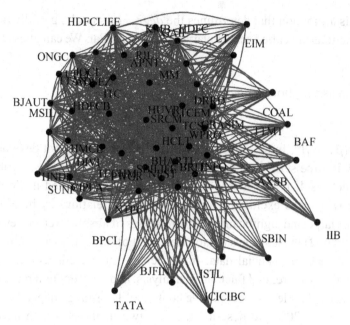

Figure 3.10 Return comovement network shown in Figure 3.4 after normalizing each return series by its corresponding estimated latent volatility series.

into the financial domain, in his 1963 article titled *The variation of certain speculative prices* (Mandelbrot, 1997). Interested readers may consult, for example, Mandelbrot (2013) and Mandelbrot and Hudson (2007) for discussions on the usefulness of fractals, self-similarity, and scaling in financial time series data (see LeBaron [2016] for an analytical discussion on Mandelbrot's legacy).

At this point it is instructive to note that the concept of scaling has been used in more than one context in financial data to refer to different phenomena. The following classification is based on the discussion in Di Matteo et al. (2005). There are two types of scaling behaviors that scientists have studied in financial data (and also in other systems). The first type is concerned with time series properties, where there is long-range correlation that decays very slowly with lags. The way such scaling behavior manifests is through empirically documented slow decay in the autocorrelation function of absolute returns. The second type is concerned with fat-tailed distribution of asset returns. As far as estimation is concerned, the first type of scaling behavior leads to a *scaling exponent* for decay whereas the second type leads to a power law for the tails of the distribution.

We can view scaling in the second sense as dilation. For a function $f(t)$ and an arbitrary constant β, the scaling property implies

$$f(\beta t) = \gamma(\beta) f(t) \tag{3.110}$$

where $\gamma(.)$ is a function that determines the *degree* of stretching or dilation. These kinds of functions are called scale invariant or self-similar. We can guess a solution:

$$f(t) = \theta t^{\phi}. \tag{3.111}$$

With such an assumption, we get

$$f(\beta t) = \theta \beta^{\phi} t^{\phi}, \tag{3.112}$$

implying $\gamma(\beta) = \beta^{\phi}$. In the remainder of the chapter we will not develop the idea of fractality in time series and we will also not discuss power law distributions in returns (which itself is considered to be a *stylized fact* as listed in Cont [2001]). Instead, we will focus exclusively on long-range correlations by building on our discussion of large and significant autocorrelations of absolute returns even at large lags. Weron (2002) present a systematic treatment of the statistical features of long-range dependence in financial data. In particular, three methods can be pursued here: R/S analysis, detrended fluctuation analysis, and periodogram regression. The third of these has had less success in the context of long-range dependence. Below, we follow Weron (2002) to describe the first two methods, which have become relatively common toolkits in such analysis. Finally, we will review the ARFIMA and FIGARCH classes of models.

3.5.1 R/S Analysis and the Hurst Exponent

Mandelbrot and Wallis (1969) came up with the technique of R/S analysis in the context of long-range statistical dependence (they called it "long run" rather than "long range," which is a more commonly used term now). The Hurst exponent, named after Harold Hurst, is the characteristic value of rescaled range analysis. Granero et al. (2008) provide a list of values of the Hurst exponent for different stochastic processes. Theoretically, white noise would have $H = 0$ and a Brownian motion would display a value of $H = 0.5$. A process with H between 0 and 0.5 would be anti-persistent, whereas if H is between 0.5 and 1 the process would be persistent. H would be equal to 1 for a trending process. While this concept has had very successful applications in fractal geometry developed by Benoit Mandelbrot (see, e.g. Mandelbrot, 2013), we will present it in an elementary fashion purely from the point of view of statistical estimation.

Let's imagine that we have a time series vector $X_T = \{x_1, x_2, \ldots, x_T\}$. First, we construct n non-overlapping subsets of equal length t_n such that $n \times t_n = T$. Let us denote the jth subset (or window) by S_j where $1 \leq j \leq n$. For the ease of description, let us denote the ith term in the original time series vector X_T corresponding to the jth window as $\tilde{x}_{i,j}$ where $1 \leq i \leq t_n$. For the jth window, we calculate the first two moments:

$$\mu_j = \frac{\sum_{i=1}^{t_n} \tilde{x}_{i,j}}{t_n},$$

$$\sigma_j = \sqrt{\frac{\sum_{i=1}^{t_n} (\tilde{x}_{i,j} - \mu_j)^2}{t_n}} \quad \text{for } 1 \le j \le n. \tag{3.113}$$

Then we find the cumulative sum of the deviations of the terms from the sample mean:

$$y_{i,j} = \sum_{k=1}^{i} (\tilde{x}_{k,j} - \mu_j). \tag{3.114}$$

Then we define a *range* term capturing the spread between the minimum and the maximum of all such deviations within the jth window:

$$R_j = \max_i \{y_{i,j}\} - \min_i \{y_{i,j}\}. \tag{3.115}$$

We find the *rescaled range* (denoted by R/S_n) of the n windows by calculating the average of the ranges of all windows scaled by the corresponding standard deviations:

$$R/S_n = \frac{1}{n} \sum_{j=1}^{n} \frac{R_j}{\sigma_j}. \tag{3.116}$$

Next, we vary the values of n: that is, the number of divisions we created in the time series. A good candidate would be to consider the powers of two: $n = 1, 2, 4, 8$, etc. Obviously, after a while, the number of data points in each window (t_n) would be so small that the above calculation would not be credible. Typically, t_n is taken to be at least 8. Then we utilize the original expression proposed by Hurst to estimate the corresponding exponent:

$$R/S_n = c \times t_n^H. \tag{3.117}$$

A straightforward approach would be to take the log on both sides and find the Hurst exponent H by linear regression with the sample analog of the left-hand side (augmenting the model with an error term):

$$\log(R/S_n) = C + \beta_H \log t_n + \epsilon_{H,n}. \tag{3.118}$$

A simple ordinary least square estimation will give the value of β_H which is the estimated value of the Hurst exponent H.

3.5.2 Detrended Fluctuation Analysis

While detrended fluctuation analysis has its origin in studying long-range correlations in DNA molecules (Peng et al., 1994), this technique is widely applicable; in fact, over the course of the previous two decades it has become a staple technique even in the context of financial data. The technique applies the same box-plot idea

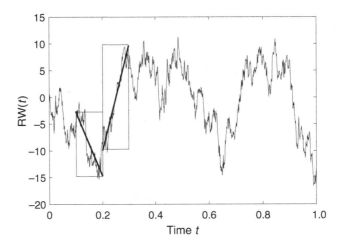

Figure 3.11 Example of converting white noise into a random walk and then fitting piece-wise linear functions.

as R/S analysis and instead of finding maxima and minima, it directly conducts a statistical estimation.

Just like the R/S analysis, we construct n non-overlapping subsets from a given time series, of equal length t_n such that $n \times t_n = T$. As we have done above, let us denote the jth subset by S_j where $1 \leq j \leq n$ and let us denote the ith term in the original time series vector X_T corresponding to the jth window as $\tilde{x}_{i,j}$ where $1 \leq i \leq t_n$. Consider a vector $Y_T = \{y_1, y_2, \ldots, y_T\}$ representing a given subset of a time series. First, we find the sample mean

$$\mu = \frac{\sum_{i=1}^{T} y_i}{T}, \tag{3.119}$$

and we create a new series based on the cumulative sum of the deviations from the mean:

$$x_t = \sum_{i=1}^{t} (y_i - \mu) \tag{3.120}$$

for time index t. This transformation converts the white noise into a random walk (see Figure 3.11).

For the jth window, we fit a straight line with the following model:

$$x_{j,t} = \alpha_j + \beta_j t + \varepsilon_t, \tag{3.121}$$

where $t = 1, \ldots, t_n$. Let us denote the line of best fit as z so that

$$z_{j,t} = \hat{\alpha}_j + \hat{\beta}_j t, \tag{3.122}$$

where $\hat{\alpha}$ and $\hat{\beta}$ denote the estimated α and β. Now, we find the *fluctuations* of the x series around this line of best fit z:

$$F_n = \sqrt{\frac{1}{T}\sum_{t=1}^{T}(x_t - z_t)^2}. \tag{3.123}$$

We compute the F_n for many values of n, and then a linear regression of the form

$$\log(F_n) = C + \beta_{dfa}\log n + \varepsilon_{dfa,n} \tag{3.124}$$

will give us the estimated value of β_{dfa} which represents the exponent found by analyzing detrended fluctuations.

3.5.3 Fractional Integration and Long Memory: ARFIMA and FIGARCH

Finally, we will discuss a generalization of the ARMA(p, q) model which can be represented as ARIMA(p, d, q), where d is the number of differences required to make the series stationary. Using notations similar to Equation 3.29, we can write

$$\alpha(B)(1 - B)^d x_t = \beta(B)\varepsilon_t. \tag{3.125}$$

In the case of the ARIMA(p, d, q) model, the order d is taken to be an integer.

Sowell (1992) considered a generalization by taking the parameter d as a fraction (see Granger and Joyeux [1980] and Hosking [1984] for prior work). The resulting process is called the *autoregressive fractionally integrated moving average* process: ARFIMA. Conceptually it lies between $I(0)$ and $I(1)$ process: that is, between a stationary and a unit root process. If d is negative, then the process exhibits antipersistence. The main empirical motivation arises from the fact that often financial time series exhibit long memory in mean or in variance. Fractionally integrated processes can nicely replicate those features.

First, we need a way to operationalize fractional d. Consider a standard binomial expansion:

$$(1 - B)^d = \sum_{j=0}^{\infty}\binom{d}{j}(-B)^j. \tag{3.126}$$

For fractional d, the infinite series serves the purpose of the lag polynomial. However, to ensure weak stationarity, we need $-0.5 < d < 0.5$. In order to see how $\binom{d}{j}$ can be computed for such values of d, we can write the above equation in the form of a gamma function which generalizes the idea of factorial to fractional numbers:

$$(1 - B)^d = \sum_{j=0}^{\infty}\frac{\Gamma(j - d)B^j}{\Gamma(-d)\Gamma(j + 1)}, \tag{3.127}$$

where the Γ function is defined as

$$\Gamma(d) = \int_0^{\infty} x^{d-1}exp(-x)dx. \tag{3.128}$$

Similar to Equation 3.125 above, we can use the same notation to write an ARFIMA model as

$$\alpha(\mathcal{B})(1 - \mathcal{B})^d x_t = \beta(\mathcal{B})\varepsilon_t, \qquad (3.129)$$

except that here d is a fraction (see Baum and Wiggins, 2001).

The autocorrelation function exhibits very slow decay and for large lags, it can be shown that the autocorrelation decays in a power law:

$$\rho_j \approx c \times j^{2d-1} \qquad (3.130)$$

for large values of j. This expression gives a crude way to estimate d by noting that a log transform of the above equation (with absolute value of ρ_j to avoid taking the log of negative numbers) will produce

$$log \, |\rho_j| \approx log \, c + (2d - 1)log \, j. \qquad (3.131)$$

Initial attempts in the literature on this topic focused on linear regression to find $\hat{d}$, that is, the estimate of d. However, this estimation is not accurate and Sowell (1992) showed that the maximum likelihood approach works better.

Next, we consider long memory in variance rather than in level. This model was introduced by Baillie et al. (1996) (see also Bollerslev and Mikkelsen, 1996). Consider the GARCH(1,1) model described in Equation 3.107. The same model can be written as an ARMA process in x_t^2 as follows:

$$x_t^2 = \omega + (\alpha + \beta)x_{t-1}^2 + \xi_t - \beta\xi_{t-1} \qquad (3.132)$$

where $\xi_t = x_t^2 - \sigma_t^2$. Similarly to the ARFIMA model, we can now generalize this process by introducing fractional differencing as follows:

$$(1 - \mathcal{B})^d x_t^2 = \omega + (\alpha + \beta)(1 - \mathcal{B})^d x_{t-1}^2 + \xi_t - \beta\xi_{t-1}. \qquad (3.133)$$

It is a challenging task to ensure positivity of conditional variance. Baillie et al. (1996, p. 22, footnote) provide conditions for a FIGARCH$(1, d, 1)$ specification. Bollerslev and Mikkelsen (1996) provide an alternate set of conditions. Conrad and Haag (2006) discuss the problem in detail and provide a more general set of conditions.

The literature on long memory has seen an amalgamation of ideas from very different origins. Graves et al. (2017) provide a concise description of the evolution of such modeling in the context of long memory starting with the work of Harold Hurst and Benoit Mandelbrot and showing how it eventually led to the econometrics literature via fractional integration models. While our examples come mostly from economic and financial contexts where long-range correlation in volatility is observable, there is a rich literature around applications of such models to biological and physical systems where similar long-range correlations are observable (see Peng et al., 1992; Peng et al., 1995; Stanley et al., 1993). Gao et al. (2006) provide some useful tips on detecting long-range correlations in time series data from a general systemic perspective.

3.6 Taking Stock and Further Reading

In this chapter, we have discussed some concepts in time series analysis, mostly from an econometric point of view. We have introduced *white noise* as the fundamental building block of a time series model. We have shown that a linear combination of white noise, in the form of an autoregressive moving average model, can capture a wide range of behavior. In particular, in order to model stationary time series, the ARMA class of models suffices thanks to the Wold decomposition theorem. Then we have discussed non-stationary time series and described statistical tests to detect the presence of non-stationarity in the data. Finally, we have extended the discussion of non-stationarity into fractional integration, which allows us to model long-range correlations. Our discussion of modeling the first moment is supplemented by a discussion of GARCH-type models that allow us to model time-varying conditional volatility.

This coverage is introductory in nature. There are many interesting and useful topics in time series analysis that we have not covered here. For example, with the toolkit we have described, one can get into a discussion on forecasting which is very useful from an econometric point of view (Elliott and Timmermann, 2016). Here we have deliberately skipped this topic as it does not directly link to the models of complex socio-economic systems, although recent advances in dynamical systems theory provide a link in a more general context (Wang et al., 2016). There are some excellent textbooks on time series econometrics that interested readers can consult: Hamilton (1994), Brockwell and Davis (2016), Neusser (2016), Enders (2008), and Tsay (2010). Some earlier attempts to discuss time series topics in the context of complex systems can be found in Mantegna and Stanley (2007) and Sinha et al. (2010).

A number of important directions have been pursued in the literature. Here we briefly describe a few of them and give some references. Autoregressive process with distributed lags (ARDL) is a well-developed toolkit for estimating dynamic relationship between multiple variables (Cho et al., 2021). Advances in modeling sampling with mixed frequencies (MIDAS) have made it possible to incorporate a variety of data with different time series frequencies in the same model for the purpose of forecasting (Ghysels et al., 2007). High-frequency financial data analysis has become a very important topic in itself (see e.g. Aït-Sahalia and Jacod,

2014). The idea of scaling that Benoit Mandelbrot introduced (Mandelbrot, 1997) has now been extended to multi-scaling, for example, by Di Matteo (2007). Finally, time series analysis has also led to a much deeper understanding of financial networks and their fragility or robustness with respect to endogenous and exogenous shocks. Diebold and Yılmaz (2015) have developed a network-theoretic way to model shock propagation across financial and economic entities, where they infer connectivity across entities by applying a vector autoregression model. In complementary developments, state space models such as Kalman filters and particle filters (and more sophisticated variants thereof) have become a very useful way to estimate agent-based models of financial dynamics that sheds light on scaling and other critical behaviors (e.g. Lux and Marchesi, 1999; Lux, 2018).

Part III

Patterns and Interlinkages

4

Pattern Recognition in Complex Systems
Machine Learning

4.1 Patterns in the Data

Machine learning as a discipline is fundamentally concerned with finding patterns in large-scale data. In recent times, it has seen enormous success in various applied fields – for example, in automated spam detection, fraud detection, and recommendation systems. The advent of computing power complemented the development of newer architectures such as distributed computing systems (Lamport, 2019). The key components in developing such architecture are scalability and the ability to handle fault tolerance in large-scale computation (Birman, 1993; Ghemawat et al., 2003).

In this chapter, we will approach the topic from a *systemic* point of view. The main idea is to find out whether discernible patterns exist in complex socio-economic data. Such datasets typically possesses the three Vs of big data: volume, velocity, and variety. As we look into the tools and techniques we will keep in mind that fundamentally the data universe will be defined by these three Vs, and therefore the techniques will also need to be flexible enough to accommodate various types of data samples. Before proceeding any further, we should note that the techniques described below are much more generally applicable and are not confined only to the socio-economic world. We emphasize socio-economic applications only in order to preserve the theme that runs throughout the book.

An important question to ask at this point is: exactly what is the relationship between the complex systems literature and the machine learning literature? The relationship of machine learning to complex systems is not widely recognized, although for quite some time many researchers have utilized machine learning toolkits to analyze complex systems. The core reason for examining methodological convergence is as follows. We have seen before that complex systems are generally characterized by the evolution of a large number of interrelated constituent parts. Thus the behavior of the system can be studied at three different levels:

first, at the macroscopic level in terms of aggregate time series and statistical behavior; second, at the mesoscopic level in terms of networks and subnetworks of the system and their behavior; and finally, at the microscopic level where the joint evolution of the dynamics of constituent entities provides the most elaborate description of the system. However, the third level naturally also possesses the most complex data structure with large numbers of entities each with potentially non-trivial dynamics. Analysis of such behavior requires two different approaches. First, we need to compress the size of the data without sacrificing too much information. Second, we need to find patterns that may not be explicit in a large volume of data. Machine learning tools are the exact fit to carry out these two tasks. They have often been useful in characterizing interrelations and patterns in the parallel dynamics of the entities being analyzed, which may not be particularly obvious if one considers the system as a whole. Therefore, we follow an atheoretical approach which attempts to find patterns without explicitly modeling each of the constituent entities separately.

As we delve into the subject matter, it is worth emphasizing here that we do not use the term "artificial intelligence" in this chapter, to deliberately de-emphasize the cognitive aspect of machine learning. There are excellent textbooks that cover the following topics with motivations arising out of the literature in computer science and cognition. We will review some of them at the end of the chapter. Readers interested in complementary developments in the machine learning literature should consult them.

4.2   Types of Learning Models

Supervised learning is an approach in machine learning where a given algorithm is trained deliberately to perform a certain task based on known observations; once the algorithm has *learned* how to do it, it is then used to make predictions given new observations. These kinds of learning algorithms are common in regression and classification tasks. Typically, the algorithm starts with a dataset for training $(D = \{(x_1, y_1) \dots, (x_n, y_n)\}, x_i \in \mathbb{R}^p)$ to perform either classification in m groups $(y_i \in \{g_1, \dots, g_m\})$ or regression $(y_i \in \mathbb{R})$. Once the algorithm is deemed suitable for making predictions, we use test data $(x_0 \in \mathbb{R}^p)$ to predict y_0 given x_0. *Unsupervised learning* algorithms are mainly for data clustering on a given set of data $(D = \{(x_1, \dots, x_n\}, x_i \in \mathbb{R}^p)$. Essentially, the end goal is to cluster data by similarity measures, and we let the data speak on the underlying relationships.

There is also *reinforcement learning*, which continually learns to perform better in a given environment. We do not discuss reinforcement learning in this chapter.

4.2.1 Bias–Variance Trade-Off

Model-building involves a trade-off. Simpler models tend to underfit any given sample dataset, but generally retain explanatory power across samples. Complicated models tend to fit any given sample dataset very well, but perform poorly when tried on other samples. In real-world applications, it is often not at first clear where to draw the line in terms of model complexity. Thus, one important concept the modeler has to consider is defined as the *bias–variance* trade-off.

Bias represents the difference between the true observed values and the values predicted by the model. A simple model may miss a lot of patterns in the data and consequently become biased, resulting in a high rate of errors in training and test data. *Variance* represents the sensitivity of the model with respect to samples. Complicated models may capture patterns in a given dataset nicely. But during the process, it may recognize even noise as a pattern. Therefore, when applied to other samples, the model performs poorly as noise is typically idiosyncratic to different samples. Therefore, such models perform well on the training data but poorly on out-of-sample data or test data.

To examine the types of error involved, let the dependent variable be denoted as Y and the independent variable as X, related to each other in the following way:

$$Y = f(X) + \epsilon, \tag{4.1}$$

where ϵ is a zero-mean error term, independent of X. We may estimate a model $\hat{f}(X)$ of $f(X)$ resulting in the expected squared prediction error at a point x as

$$total\ error(x) = E[(y - \hat{f}(x))^2]. \tag{4.2}$$

With a little bit of algebra, one can show that this error term can be decomposed into three components:

$$total\ error(x) = (f(x) - E[\hat{f}(x)])^2 + E[(\hat{f}(x) - E[\hat{f}(x)])^2] + \sigma_\epsilon^2. \tag{4.3}$$

These three components are called bias, variance, and irreducible error. The last component arises because there will always be some variation in the data that cannot be explained by covariates. In Figure 4.1, we show four panels. Panel (a) illustrates the concepts of underfitting and overfitting. Panels (b), (c), and (d) show three different curves fitted on the same dataset, with different levels of bias and variance.

Some of the models we will describe below tend to suffer from underfitting and some suffer from overfitting. For example, linear and logistic regression models often underfit the data. On the other hand, models such as decision trees, support vector machines, and neural networks often suffer from overfitting. Therefore, one has to be careful in designing a model and making predictions from it, knowing the relative limitations and strengths of different modeling approaches.

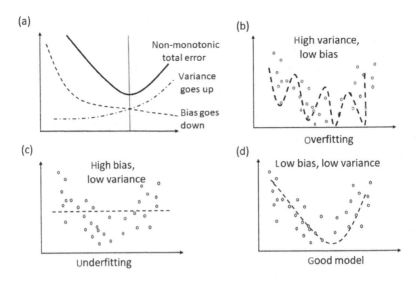

Figure 4.1 Schematic representation of (a) bias–variance trade-off, (b) overfitting, (c) underfitting, and (d) best fit. In panels (b), (c), and (d), the dashed lines represents fitted curves and the circular points are identically distributed.

4.3 Modeling Dependence via Regression

Regression analysis, one of the most important statistical and econometric toolkits, has also found its place as a starting point in machine learning. It starts with a set of independent variables, assesses how dependent variables are related to independent variables, and then proceeds to make inferences or forecasting depending on the modeler's choice. Interestingly, the least square method of finding the best fit curve has its origin in physics. Carl Friedrich Gauss, the legendary mathematician, investigated the method of least squares in the context of planetary motions as early as 1794 (Gauss, 1809). Adrien Legendre, another great mathematician from the same era, also worked on the method of least square estimation (Legendre, 1805). This was the first systematic treatment of curve-fitting by minimizing the squared residuals between the observations and the model's predicted values. Today it has become a standard tool that cuts across empirical analysis in all fields, ranging from economics to political science to physics and biology. Below we describe some of the best-known regression models.

4.3.1 A Numerical Look at Linear Regression

We described the linear regression model in Section 2.3.5. It will be useful to revisit it in the context of machine learning so that it can serve as a building block for the other topics. Using the standard notation, we write a p-variable regression model for n observations as

$$y_i = \beta_0 + \beta_1 x_{1i} + \beta_2 x_{2i} + \cdots + \beta_p x_{pi} + \epsilon_i$$

for $i = 1, \ldots, n$. The coefficient β_j represents the slope of y with respect to variable x_j. The final term ϵ_i represents the error term of the residual. We will not repeat the complete description that we gave in Section 2.3.5. Instead, we will elucidate a number of concepts that have become very useful in the machine learning literature in general. We have already seen that in matrix form, the model is written as

$$Y = X\beta + \epsilon \tag{4.4}$$

and the least square estimate aims to minimize the sum of the squared residuals

$$\epsilon'\epsilon = (y - X\beta)'(y - X\beta). \tag{4.5}$$

To recap the result, the optimal choice of β is given by

$$\hat{\beta} = (X'X)^{-1}(X'Y). \tag{4.6}$$

Therefore, the predicted value would be given by

$$\hat{Y} = X\hat{\beta}. \tag{4.7}$$

With this formulation, let us now define some terms which capture the numerical aspects of finding the $\hat{\beta}$.

Cost Function

Models like linear regression come with a cost function. This cost appears in the form of how bad the model is at predicting the dependent variable. Therefore, the goal of the modeler would be to minimize the cost of the model, which in turn makes a model better in terms of predictions.

One standard way to define a cost function here is to consider the residual sum of squares (sometimes also called *sum of squared residuals*) which measures the discrepancy between data and the values predicted by the model. Mathematically, it can be written as

$$RSS = \sum_{i=1}^{n} (y_i - \hat{y}_i)^2. \tag{4.8}$$

Here, $\hat{y}$ is the predicted value given in Equation 4.7. Mean squared error (MSE) measures the average of the squares of the errors:

$$\text{MSE} = \frac{1}{N} RSS \tag{4.9}$$

$$= \frac{1}{N} \sum_{i=1}^{n} (y_i - \hat{y}_i)^2. \tag{4.10}$$

Clearly, a model with smaller RSS (or MSE, for that matter) would be a better model. So the next question is how to minimize the RSS numerically?

Cost Minimization

In Equation 4.7, we have already seen the theoretical solution of the minimization problem. But in many cases, the solution may not be analytically tractable and one has to utilize numerical methods. Therefore, here we review a numerical approach to find the minimum on a surface. For simplicity, let us consider a univariate analog of Equation 4.7 with only two parameters, β_0 and β_1.

The goal is to minimize the sum of the squared errors denoted by a function f, which is a function of the free parameters β_0 and β_1. Thus, the function can be described as $f(\beta_0, \beta_1)$. The idea of the *gradient descent* technique is to find the gradient on the surface and continually move opposite to the gradient, to reach a local minima. The gradient is written as

$$\nabla f(\beta_0, \beta_1) = \begin{bmatrix} \frac{df}{d\beta_0} \\ \frac{df}{d\beta_1} \end{bmatrix}. \tag{4.11}$$

The gradient at any given point gives the slope of the cost function and the direction we should move to update our parameters. One needs the $f(.)$ function to be continuous and continuously differentiable to apply this technique. There are other techniques which are applicable for functions which may not have derivatives at all points. We will not elaborate on the numerical techniques here. Interested readers can consult textbooks on numerical optimization for a complete discussion (see, e.g. Nocedal and Wright, 2006).

Explanatory Power of a Model

R^2 or the coefficient of determination is a measure of the explanatory power of a model. Specifically, it expresses how much of the variation in Y is explained by the variation in X. It is expressed as

$$R^2 = 1 - \frac{\text{RSS}}{\text{SS}_{total}}$$

$$= 1 - \frac{\sum_{i=1}^{n}(y_i - \hat{y}_i)^2}{\sum_{i=1}^{n}(y_i - \overline{y}_i)^2}. \tag{4.12}$$

In the above, SS_{total} stands for total sum of squares, indicating the variance in the data, and $\overline{y}_i$ represents the average. Due to the normalization, the value of R^2 always lies between 0 and 1, where 0 represents no explanatory power and 1 represents full explanatory power of the model.

As a measure, R^2 has a problem. If one increases the number of variables even without any regard to whether it really is explanatory or not, R^2 cannot decrease.

Therefore, a high R^2 for a model may actually be driven by many unimportant variables plugged into the model. In order to tackle this problem, one adjusts the R^2 value by the number of explanatory variables such that a higher number of explanatory variables come with a penalty term. It is expressed as

$$R^2_{adj} = 1 - \left[\frac{(1 - R^2)(n - 1)}{(n - p - 1)} \right], \tag{4.13}$$

where p is the number of independent regressors. Here it is useful to note that we have previously seen this concept in our discussion on statistics (Section 2.3.7).

4.3.2 Response Variable in Logistic Form

Logistic regression is a common form of supervised classification algorithm. It is heavily used in econometrics and statistics (see Section 2.3.6) although the approach and interpretation differ slightly in that in econometrics, typically inference (Greene, 2003) plays a bigger role than in the classification exercise *per se*. Here we will present it as a tool to classify observations probabilistically based on covariates (Nelder and Wedderburn, 1972).

Let's consider a response variable y which can take only discrete values. The variable is often also described as a "target" variable and it is a function of the underlying X with p covariates. These underlying variables are sometimes called as "features" in the context of machine learning. But for our purpose, we can simply treat them as covariates as we considered them in our earlier discussion in Chapter 2.

We use a *logistic* regression to predict a binary outcome (in the sense of 1 / 0 for y) given a set of independent variables X. The key point is that the model will construct a probability of a particular event happening (say $y = 1$) as a function of X. This probability is called log-odds and we will denote it in the following by the notation $\mathcal{P}$.

Let us start with a baseline specification linking the log-odds ratio and the covariates:

$$\log \mathcal{P} = f \left(\sum_i^p \beta_i x_i \right) \tag{4.14}$$

where $f(.)$ is an increasing function with respect to its argument. The goal is to estimate the optimal parameter vector $\hat{\beta}$. Here we assume that one of the covariates can be a vector of ones (i.e. the above model may contain a constant term). Given that $\mathcal{P}$ has to be between 0 and 1, we normalize it using the functional form of a logistic equation:

$$\mathcal{P} = \frac{exp(\sum_i^p \beta_i x_i)}{1 + exp(\sum_i^p \beta_i x_i)}. \tag{4.15}$$

It is the use of this functional form that gives rise to the name of logistic regression. Denoting the linear combination of covariates by z, we can write $z = \sum_i^p \beta_i x_i$. Therefore, we get

$$P = \frac{exp(z)}{1 + exp(z)} \tag{4.16}$$

which can also be written as

$$z = \log\left(\frac{P}{1 - P}\right). \tag{4.17}$$

One can use maximum likelihood estimation to find the optimal parameter vector $\hat{\beta}$. Note that the logistic regression method by itself is not a classifier. But it can be easily made into a classifier by choosing a cut-off for the variable P. The above two-outcome model can be generalized to multinomial logistic regression to handle multiple outcomes. Given this reason, it could have also been described as part of the discussion on classification algorithms below.

4.3.3 Shrinking Coefficients

There are some important estimation drawbacks of the linear models. For example, for estimating β parameter we need to calculate the term $(X'X)^{-1}$, which is impossible if the data matrix X (or the product matrix $(X'X)$, for that matter) does not have full rank. This situation occurs when the covariates or predictors X show multicollinearity. Furthermore, for a large number of predictors the traditional estimation for β often lacks the ability to accurately identify the true parameters. Regularization methods are used to overcome these problems of parameter estimation. In the following, we discuss a few well-known techniques from this literature. The essential idea is to shrink the parameters so that unless they are statistically useful for explanatory purposes, the model shrinks their corresponding magnitudes, bringing them close to zero. This idea follows from the pioneering work by Tibshirani (1996) among others.

Method of Penalized Least Square

In penalized least square estimation, we estimate the regression model parameters by minimizing the residual sum of squares with a pre-defined penalty. Let λ be a tuning parameter that controls the shrinkage and let $f(\beta)$ be a penalty term. We obtain the estimates of β by minimizing the following equation:

$$\hat{\beta}_{pls} = \arg\min_{\beta} ||Y - X\beta||^2 + \lambda f(\beta) \tag{4.18}$$

with the tuning parameter $\lambda \geq 0$. Clearly, when $\lambda = 0$, the above equation becomes an ordinary least square estimation problem.

Ridge Regression

An alternative to the penalized least square is the ridge estimation (Hoerl and Kennard, 1970), where we use an L–2 penalty, that is, $f(\beta) = ||\beta||^2$. Hence, we define the ridge estimate for β:

$$\hat{\beta}_{ridge} = \arg\min_{\beta} ||Y - X\beta||^2 + \lambda ||\beta||^2, \quad \lambda \geq 0. \tag{4.19}$$

This yields the solution for β parameters

$$\hat{\beta}_{ridge} = (X'X + \lambda \mathbf{I})^{-1} X'Y \tag{4.20}$$

where $\mathbf{I}$ is a $p \times p$ identity matrix. The addition of $\lambda \mathbf{I}$ solves the issue of inverting the term $(X'X + \lambda \mathbf{I})$ in the presence of multicollinearity and provides a unique solution to the estimates. It can be shown that while the ridge estimator is biased, it provides stable estimates of the regression model parameter β.

Least Absolute Shrinkage and Selection

The least absolute shrinkage and selection operator (LASSO) includes an L–1 penalty, that is, $f(\beta) = ||\beta||$ in the parameter estimation (Tibshirani, 1996). This estimation procedure is often important as the ridge method does not provide an easily interpretable model due to the nature of not selecting important predictors for the model. For the use of L–1 penalty, LASSO becomes a convex problem and hence solves the equation efficiently. In addition, for the sparse solution approach, LASSO has predictor selection properties. We write the equation that we want to minimize to estimate β as

$$\hat{\beta}_{lasso} = \arg\min_{\beta} ||Y - X\beta||^2 + \lambda ||\beta||, \quad \lambda \geq 0. \tag{4.21}$$

Solving the above equation has the advantage of variable selection for the linear regression model and also provides an interpretable estimate of β, even in situations when the number of predictors is more than the number of observations, that is, $p > n$. However, LASSO has the drawback of getting accurate predictions compared to the ridge regression when $n > p$. Furthermore, LASSO, tends to selects arbitrary predictors from a highly correlated group. This problem can be ameliorated by group LASSO. Zou (2006) proposed a modification in the form of adaptive LASSO, which has the oracle property: that is, the estimator performs as if the true model was known in advance.

Elastic Net

LASSO selects maximum n predictors when $p > n$, that is, when the number of predictors is greater than the number of observations. Zou and Hastie (2003) proposed a combination of L–1 and L–2 penalties to estimate the model parameters β

and called it elastic net. Let λ_1 and λ_2 be tuning parameters and $\lambda_1, \lambda_2 \geq 0$. Hence, we write the elastic net estimator for β by minimizing the following equation:

$$\hat{\beta}_{net} = \arg\min_{\beta} ||Y - X\beta||^2 + \lambda_1 ||\beta||^2 + \lambda_2 ||\beta||, \quad \lambda_1, \lambda_2 \geq 0. \qquad (4.22)$$

The penalty terms can be redefined as

$$\alpha ||\beta||^2 + (1 - \alpha)||\beta|| \qquad (4.23)$$

where $\alpha = \frac{\lambda_2}{\lambda_1 + \lambda_2}$.

A general feature of all of these methods is that they penalize unnecessary parameters and this penalty acts by shrinking them toward zero. These methods have geometric interpretations as well, which we have completely omitted discussing here. Interested readers can consult James et al. (2013) for discussion.

4.4 Low-Dimensional Projection

It is tempting to think that more data are always beneficial. In a sense, that is true. But it may come with real costs. Data with a large number of variables are generally more difficult to visualize and analyze. Imagine dealing with a dataset with thousands of variables. While such data would provide substantial richness to capture all nuances of the underlying system, it may not be clear how to deal with such a large number of variables. Even more so, it may not be evident *a priori* whether considering all of the different variables will convey useful information or not. If the information content is similar across these variables, then the modeler can safely ignore a subset of these variables as they may be redundant and work with the rest so as to minimize computational cost. Therefore, while datasets with rich coverage of variables might be good for the purpose of not missing out on important information, one also has to find out which variables do actually convey most of the information and choose them to work with. With this motivation, a major component of data analysis comes in the form of low-dimensional representation of high-dimensional data such that most of the information is still preserved. Clearly this represents a trade-off. Lower-dimensional data make computation easy but also necessarily lead to some loss of information. Thus the question is how to quantify that trade-off and how to decide on which subset of variables (or combinations thereof) to select. Below, we describe three important techniques developed in the literature: principal component analysis, factor analysis, and linear discriminant analysis. Additionally, we also present a fourth technique called multidimensional scaling which has a different flavor from the other three.

Before we start discussing the details, one small point should be noted. In the following, we have denoted a dataset by a matrix $X_{p \times n}$ where the p denotes the number of variables and n denotes the number of observations of these variables.

This might be somewhat confusing because in the case of panel data, many books and papers use the notation $X_{n \times t}$ where n is the number of variables and t is the number of observations (time points). In fact, we ourselves have used this notation in Chapter 3 (e.g. while describing the vector autoregression model in Equation 3.44). In this chapter, we will stick to the usage of n as the number of observations as that is more consistent with the notations typically used in this literature.

4.4.1 Principal Component Analysis

Principal component analysis is a well-known dimension reduction technique. The idea is to project the original data on to a smaller dimensional space while preserving the variation in the data as much as possible. Nowadays, principal component analysis is extensively used across a variety of fields, from econometrics to engineering to neuroscience to physics. But this technique is not new. In fact, it has been known for more than a hundred years now (Pearson, 1901; Hotelling, 1933). The ease of computation along with the intuitive interpretation of the result has made the technique widely applicable and popular (Alpaydin, 2020).

Suppose we are given a p-variable dataset with n number of observations: $X = \{x_1, x_2, \ldots, x_n\} \in \mathbb{R}^{p \times n}$ with $p << n$, where each x_i is a column vector of size $p \times 1$. Let us assume that the data pertaining to each variable is normalized by mean. If the data is unnormalized, we can normalize the raw data y to $x = y - \langle y \rangle$ where $\langle . \rangle$ represents the sample average for a vector. The idea is to find a smaller number of variables (say, $k < p$) which preserves most of the variation in the data, given by the empirical covariance matrix

$$S = \frac{1}{n} X X'. \tag{4.24}$$

In order to do that, let us create weight vectors $\{v_j\}$ where $j = 1, \ldots, k$ such that each weight vector is of size $p \times 1$ and has unit norm. Let us pick up the first weight vector and create a projection:

$$y_1 = X' v_1 \tag{4.25}$$

which is a vector of size $n \times 1$. Let us denote the jth row of the matrix X' by $X'_{(j)}$. If the weight vector v_1 maximizes variance, then it is the solution of the optimization problem

$$v_1 = \underset{v}{\mathrm{argmax}} \sum_{i=j}^{n} \left((X'_{(j)} . v)^2 \right). \tag{4.26}$$

The next vector v_2 has to be orthogonal to v_1 and has to maximize the residual variance. The next vector v_3 has to be orthogonal to v_1 and v_2 and has to maximize the residual variance, and so forth. PCA is based on the solution to this problem

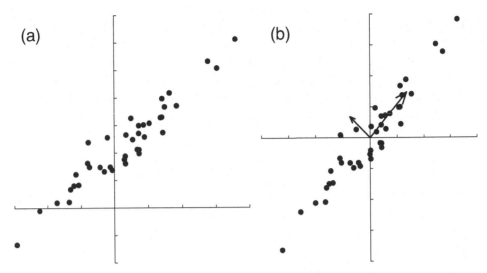

Figure 4.2 Example of PCA: (a) bivariate data with a non-diagonal covariance matrix; (b) centered data with the arrows indicating the directions of decomposition due to the eigenvectors of the covariance matrix.

of recursively finding out such weight vectors. It can be shown that if we simply conduct an eigenvalue decomposition of the covariance matrix,

$$S = \sum_{i=1}^{p} \lambda_i e_i e_i',$$

(4.27)

then the eigenvectors (e) corresponding to rank-ordered eigenvalues (λ) solve the optimization problem described above. This finding reduces the computational cost dramatically as one can bypass the whole optimization procedure and simply take up the eigenvectors corresponding to the kth eigenvalue as the kth weight vector. Figure 4.2 provides a pictorial representation of the technique in two dimensions. Sometimes, the eigendecomposition is also conducted on the sample correlation matrix rather than the covariance matrix. Finally, the explanatory power of the top k principal components is given by the ratio of the top k eigenvalues:

$$\frac{\sum_{i=1}^{k} \lambda_i}{\sum_{j=1}^{p} \lambda_j}.$$

(4.28)

The usefulness of the algorithm arises from the fact that often only the top few principal components suffice to explain a very substantial portion of the total variation in the data, even in the presence of a large number of variables. Jolliffe and Cadima (2016) present an extensive treatment of principal component analysis and many variations of it. This is a very standard technique and widely applied in financial data. Interested readers may also consult Tsay (2010) for a discussion in the context of financial time series.

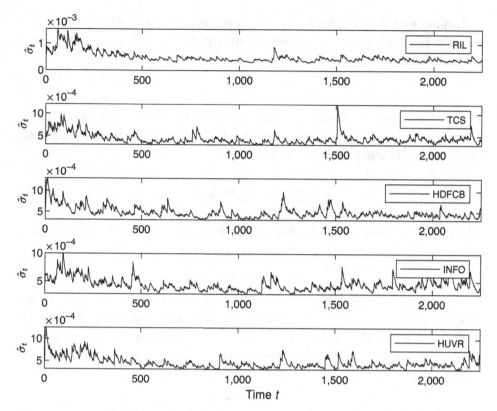

Figure 4.3 GARCH estimated latent volatility series for five companies for intra-day data. These companies feature in the network shown in Figure 3.4. We use these time series to apply the principal component analysis and factor model.

Consider the time series shown in Figure 4.3. These are volatility series estimated by the GARCH model described in Section 3.4 on five different companies listed in the Indian stock market. We apply principal component analysis on this dataset to extract the first dominant component. We show the first component in Figure 4.4. We should mention two caveats here. First, these volatility estimates are point estimates. We have not considered the error bars around these point estimates. Second, one should ensure stationarity before applying the technique. We show a factor model as well for the same time series (Section 4.4.2). There are more formal ways to model the common component of volatility in the literature: see Hu and Tsay (2014).

4.4.2 *Factor Model*

Factor modeling is built on an idea very close to that of principal component analysis (Alpaydin, 2020). It also attempts to capture the covariance matrix as much as possible by modeling the data via a set of underlying *latent factors*.

Suppose we have a dataset X of p observable variables with n observations. For the sake of simplicity, let us denote the ith variable as $\{x_{il}\}_{l=1,\ldots,\,n}$. The sample covariance matrix is given by

$$S = \frac{1}{n}XX'. \tag{4.29}$$

In factor analysis, the modeler assumes the existence of k number of factors f_j (where $j = 1,\ldots,k$) such that each observable variable can be modeled as a linear combination of these factors and an error term:

$$x_{il} = \mu_i + \sum_{j=1}^{k} w_{ij}f_{jl} + \epsilon_{il} \tag{4.30}$$

where w is the factor weight/loading matrix. In matrix notation, we can write this as

$$x_l - \mu = wf_l + \epsilon_l. \tag{4.31}$$

Now, we can make three assumptions to simplify the calculations. First, these factors have zero mean, that is, $E(f) = 0$. Second, the covariance matrix of the factors is given by an identity matrix, that is, $Cov(f) = I$. Finally, the factors are uncorrelated to the error terms. Therefore, equating the covariance matrices from Equation 4.31, we get

$$\Sigma_x = ww' + \Sigma_\epsilon \tag{4.32}$$

where Σ_x denotes the covariance matrix of x and Σ_ϵ denotes the covariance matrix of the error term (we have also used the second assumption $Cov(f) = I$ to get rid of the covariance matrix of the factors). If one makes parametric assumptions (e.g. that the data-generating process is Gaussian), then one can use maximum likelihood estimation for the parameters under the restriction given by Equation 4.32.

Factor analysis helps with both explanatory and confirmatory exploration of data. In the finance literature, many factors have been proposed that explain variations in asset returns (see the classic work on financial factor models – Fama and French, 1993). In case of such observed factors, one can estimate the factor loadings by estimating the model given by Equation 4.30. See Tsay (2010) for a detailed treatment of the topic in the context of financial time series.

We consider the same five time series as shown in Figure 4.3 and extract a latent factor by considering a single-factor model estimated with maximum likelihood. The resulting factor is shown in Figure 4.4.

4.4.3   *Linear Discriminant Analysis*

Linear discriminant analysis is another projection technique that allows dimension reduction. Here, the underlying idea is that the data is *a priori* classified into more

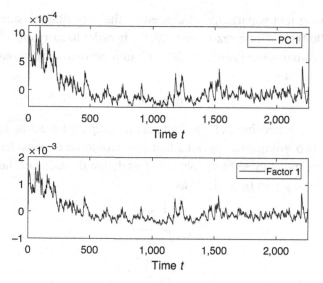

Figure 4.4 First principal component (top panel) and the latent factor obtained from a single-factor model (bottom panel). We have scaled the data by the same means as in Figure 4.3 before applying the techniques. The two series exhibit similar patterns with minute numerical differences.

than one group. The goal is to find a particular projection of the data so that the data points belonging to the same group are projected close to each other while maximizing the inter-group distance for data points belonging to different groups. The earliest analysis of linear discriminants is attributed to Fisher, who first introduced the concept and also provided a full-blown solution for the two-group case (Fisher, 1936).Duda and Hart (1973) presented a textbook treatment of this technique along with applications in pattern classifications and made it accessible to a larger audience across disciplines. Over the last few decades, this technique has been extensively used in multivariate statistics, pattern recognition, and machine learning applications, among others. Here we present a standard treatment for two-group cases. Further generalizations are possible, both in the number of groups (Duda et al., 1973) and in the domain of nonlinear methods for dimension reduction. In the current discussion, we will omit them and focus on the case with two groups.

Fundamentally, the objective of this technique is to linearly project high-dimensional data into a line by maximizing the linear separability between groups where such groups are assigned to the data points *a priori*. Let us denote the dataset being given a matrix X of size $p \times n$ where p is the number of variables and n denotes the number of observations. It might be easier if we think of the variables as features. The task is to segregate n observations (each having p number of features) into multiple groups. Put differently, the dataset can be described as a collection of vectors: $x_1, x_2, \ldots, x_n$. To begin with, we assume that there are two groups G_1 and G_2 with n_1 and n_2 number of data points within them, respectively.

The following text is partially influenced by the expositions made by Elhabian and Farag (2009) and Gutierrez-Osuna (2020). In order to carry out a linear projection, let us consider some vector $v \in \mathbb{R}^p$ with unit norm and carry out a projection of a given point x_i as

$$y_i = v'x_i \tag{4.33}$$

for $i = 1, \ldots, n$. Since the goal is to find maximal separation across the data points belonging to two groups, the obvious first step would be to consider the distance between the means of these two groups. Let us define the mean of the transformed values for the gth group ($g = 1, 2$) as

$$
\begin{aligned}
\mu_g^y &= \frac{1}{n_g} \sum_{j \in G_g} y_j \\
&= \frac{1}{n_g} \sum_{j \in G_g} v'x_j \\
&= v'\mu_g^x
\end{aligned} \tag{4.34}
$$

where μ_g^x is the mean of the data points in the gth group. Therefore, one can pick up the quantity $|\mu_1^y - \mu_2^y| = |v'(\mu_1^x - \mu_2^x)|$ to be maximized by appropriately choosing v. The problem with this approach is that the above quantity simply characterizes the difference between the mean vectors and does not account for variance. In case the variances are large within the groups, simply accounting for distance between the means would not be a good solution to discern the groups, as they might suffer from overlaps. Therefore, one also needs to account for the effects of within-group variances.

We define the within-group variances by

$$S_{yg}^2 = \sum_{j \in G_g} (y_j - \mu_g^y)^2 \tag{4.35}$$

for $g = 1, 2$. The total within-group variance is given by $S_{y1}^2 + S_{y2}^2$. Combining the above terms, we get the function of v to optimize:

$$F(v) \equiv \frac{|v'(\mu_1^x - \mu_2^x)|^2}{S_{y1}^2 + S_{y2}^2} \tag{4.36}$$

with respect to vector $v \in \mathbb{R}^p$.

We will skip the complete derivation of the solution here and will provide the solution directly for the above optimization problem. For that purpose, we need to define two quantities. The first one is a *between-group* scatter matrix in the form of

$$S_{BG} = (\mu_1^x - \mu_2^x)(\mu_1^x - \mu_2^x)'. \tag{4.37}$$

The second one is a *within-group* scatter matrix in the form of

$$S_{Wg} = \sum_{x_j \in G_g} (x_j - \mu_g^x)(x_j - \mu_g^x)' \tag{4.38}$$

for $g = 1, 2$. Now, we can define a total within-group scatter matrix as

$$S_{WG} = S_{W1} + S_{W2}. \tag{4.39}$$

Then the optimization problem resulting from Equation 4.36 can be written as

$$\max_{v} \frac{v'S_{BG}v}{v'S_{WG}v}. \tag{4.40}$$

The solution for the optimal projection vector v^* solving Equation 4.40 is given by

$$v^* = S_{WG}^{-1}(\mu_1^x - \mu_2^x). \tag{4.41}$$

See Hastie et al. (2009) for a more elaborate discussion and further extensions.

4.4.4 Multidimensional Scaling

Multidimensional scaling is a nonlinear approach to dimension reduction. In the context of data-driven analytics, it is more of a visualization tool than a mathematical representation *per se*. It is often used when a large number of entities are present in the dataset and there is a unique pairwise distance for all the entities. If one collects all the distances in a matrix, the resulting matrix is called the *dissimilarity* matrix, which encapsulates all the distances in the form of dissimilarities. If the number of entities is very small, it may be easy to eyeball the matrix to see if some groups of entities have smaller distances among them (indicating higher similarities) than with respect to the rest of the entities. However, if there is a large number of entities, it is difficult to find if there are subsets of entities such that the within-subset dissimilarities are less than across subsets. Effectively, the idea is that for a large dimensional distance matrix, it can be prohibitively difficult or even nearly impossible, simply by visual inspection, to find *subsets* which are similar within themselves but dissimilar across subsets. In cases like this, multidimensional scaling becomes extremely useful. It attempts to find a lower dimensional representation (typically on a two- or three-dimensional plot) from a given dissimilarity matrix. Such low-dimensional representation often makes the subsets clear, whether this is evident simply by eyeballing the resulting projections or more formally by applying clustering techniques.

Here it is noteworthy that such subset detection is reminiscent of community detection in networks. Indeed, at the end of this section we will discuss an example of how multidimensional scaling can be used for community detection in financial comovement networks. It is easiest to explain the concept by using an example. Probably the most intuitive example is that of p cities being the entities and the pairwise geographic distance between them being the dissimilarities (see Figure 4.5). Let us denote the dissimilarity matrix by $\mathbb{D}_{p \times p} = [[d_{ij}]]$. In the context of a two-dimensional representation, the goal is to find the coordinates $\{x_k, y_k\}_{k=1, \ldots, p}$ such that each pairwise distance d_{ij} equals (or at least approximates) the Euclidean distance between the coordinates of the nodes i and j, that is, (x_i, y_i) and (x_j, y_j).

The concept of *classical* multidimensional scaling is straightforward. Below we will describe the technique following work by Warren Torgerson and Joseph Kruskal. For further discussion on the nuances of the method, interested readers can refer to the original work presented in Kruskal (1964a), Kruskal (1964b), Torgerson (1965), and Kruskal (1978). More recent treatments can be found in Cox and Cox (2008), and Alpaydin (2020). Let us continue with the two-dimensional case to keep things simple. The key statement for the main problem is as follows. Given the matrix $\mathbb{D}$, the objective is to find coordinates that minimize the sum of the squared residual of the dissimilarity and the model-implied distance, given by

$$\epsilon^2 = \sum_{i \neq j=1, \ldots, p} \left(d_{i,j} - \|o_i - o_j\| \right)^2 \tag{4.42}$$

or some monotone transformation of the residual ϵ, where o_i represents the coordinates of the ith data point. Note that if the mapping is perfect (assuming Euclidean distance) with $\epsilon^2 = 0$, then we should get

$$d_{ij}^2 = (x_i - x_j)^2 + (y_i - y_j)^2. \tag{4.43}$$

Let us denote the corresponding matrix as $\hat{\mathbb{D}}$ (matrix with all elements squared in the given dissimilarity matrix $\mathbb{D}$). If Equation 4.43 holds, then the resulting matrix $\hat{\mathbb{D}}$ will be in the form of ZZ' for some matrix Z.

The multidimensional scaling technique uses two algebraic tricks. First, a Gram matrix G can be decomposed as a product of two matrices as $G = ZZ'$ (see e.g. Boyd and Vandenberghe [2018] for the definition and properties of a Gram matrix). Second, the matrix $\hat{\mathbb{D}}$ can be converted into a Gram matrix by double centering: $G_{\mathbb{D}} = -W\hat{\mathbb{D}}W/2$ where the centering matrix is given by $W = I - \frac{1}{p} 1_{p \times p}$ (I being an identity matrix). So the idea is to first make the dissimilarity matrix into a Gram matrix and then decompose the Gram matrix into the coordinate matrix given by Z. Once these two steps have been carried out, the remaining step is to find this coordinate matrix Z. This can be done by considering an eigendecomposition of the matrix $G_{\mathbb{D}}$ given by

$$G_{\mathbb{D}} = \sum_{i=1}^{p} \lambda_i e_i e_i'. \tag{4.44}$$

In matrix notation, we write this as

$$G_{\mathbb{D}} = e \lambda e'. \tag{4.45}$$

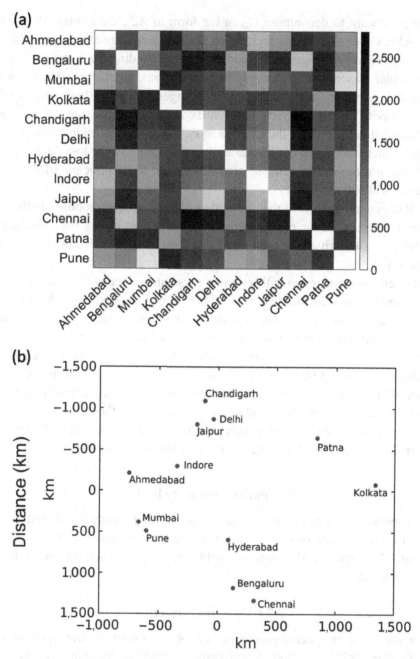

Figure 4.5 MDS map of 12 Indian cities. (a) Intercity distances (in km) among 12 Indian cities: the gradient bar shows the scale. (b) Multidimensional scaling map (two-dimensional) corresponding to those 12 cities. We have rotated the map and centered it to represent the geographical layout. It can be checked with respect to a conventional geographical map of India that the MDS map captures the actual spatial distribution of these cities reasonably well.

Since we want to decompose $G_\mathbb{D}$ in the form of ZZ', the solution is given by $Z = e\lambda^{1/2}$. Given that we are focused on the two-dimensional case, we will consider only the top two eigenvalues and their corresponding eigenvectors. There is no particular reason to consider only two dimensions, apart from ease of visualization. In general, one can consider the top k eigenvalues and their corresponding eigenvectors to do the analysis. This matrix Z serves as the dimension-reduced coordinates. In cases where the data has more than two dimensions to begin with, the same calculations will not hold exactly in the sense that the loss function given by ϵ^2 will not be exactly zero, in general. But the same principle of dimension reduction will still work.

Further generalizations of the above algorithm have been proposed in the form of metric and non-metric multidimensional scaling that allows the consideration of intangible similarities (or dissimilarities) such as similarities in colors, faces, letters and so on.

Let us end this discussion with a small digression on financial markets. In recent times, there has been considerable interest in community detection in financial markets. One approach has been to consider the pairwise correlation between asset returns to represent similarity in asset return dynamics. In this way, a market can be represented by a dissimilarity matrix where $d_{ij} = 1 - C_{ij}$, C_{ij} being the pairwise correlation coefficient between the returns of the ith and the jth assets. Once such a $\mathbb{D}$ matrix is created, one can use a multidimensional scaling technique to project it on a low-dimensional space and identify clusters. We show one such example in Figure 4.6 based on the asset return data for 115 firms listed in an Indian stock market: the Bombay Stock Exchange (see also Sharma et al., 2017).

4.5 Finding Similarity in Data

In this section, we focus on finding structures in multivariate data based on similarity (or the opposite of it, dissimilarity). The main objective is to find patterns without *ex ante* specification of what would be the expected pattern. We let the data speak for itself.

4.5.1 Spectral Projection

Clustering based on spectral projection is a robust non-parametric technique (see, e.g., Fiedler [1973] and Donath and Hoffman [2003]). Its attractiveness comes partly from a solid theoretical underpinning and partly from its ability to deal with large datasets quite easily.

Let us start with a similarity matrix W of size $p \times p$ that captures pairwise similarities across p different entities (a dissimilarity matrix can also be considered with appropriate modifications of the following steps). In case of multivariate data,

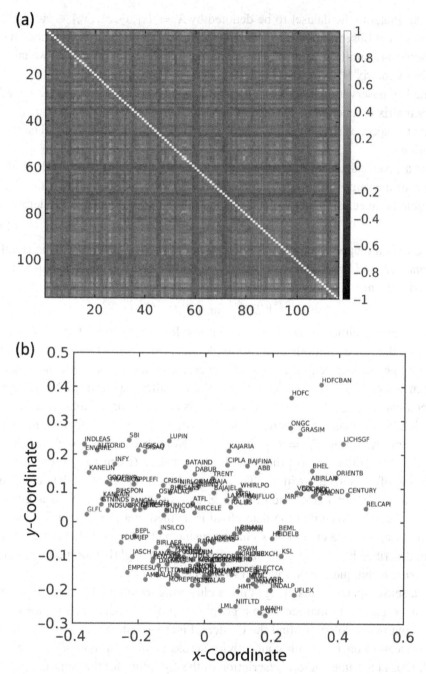

Figure 4.6 MDS map of 115 stocks traded on the Bombay Stock Exchange (BSE). (a) Correlation matrix of 115 stocks. The gradient bar represents the correlation values ranging from −1 to 1. (b) Multidimensional scaling map (two-dimensional) of the same set of 115 stocks.

one can imagine the dataset to be denoted by $X = \{x_1, x_2, \ldots, x_n\} \in \mathbb{R}^{p \times n}$ which consists of n observations on p variables. The corresponding correlation matrix can serve as a similarity matrix. For the purpose of describing the idea, we take the similarity matrix W to be given.

The key observation here is that the similarity matrix can be treated as the adjacency matrix of a network. This is not only for figurative purposes: we can actually utilize the spectral structure of the network to carry out the clustering. The idea is as follows.

For a given adjacency matrix W, we can define its Laplacian which is also a matrix of the same size (graph Laplacian is also used for spectral partitioning for community detection in networks; see Section 5.7.1). The Laplacian is defined as

$$\mathcal{L} = D - W \tag{4.46}$$

where D is a diagonal matrix with degree connectivity ($D_{ii} = d_i$). This is called an *unnormalized* Laplacian. Alternatively, a normalized version is also used in the following form:

$$\mathcal{L} = D^{-1/2} W D^{-1/2} \text{ or } I - D^{-1/2} W D^{-1/2}. \tag{4.47}$$

One can utilize either of the above definitions: for the purpose of spectral decomposition, the two definitions give rise to identical eigenvectors (Rohe et al., 2011). Once we get the Laplacian $\mathcal{L}$, we can conduct an eigendecomposition and pick only s eigenvectors $e_1, \ldots, e_s$ corresponding to the s smallest eigenvalues. Next, we construct a matrix $\mathcal{E} \in \mathbb{R}^{p \times s}$ which has the vectors $e_1, \ldots, e_s$ as columns (stacked side by side). We normalize the matrix $\mathcal{E}$ in the form of $\mathcal{Z} \in \mathbb{R}^{p \times s}$ so that the rows have unit norm, by setting $\mathcal{Z}_{ij} = e_{ij} / \left(\sum_k e_{ik}^2 \right)^{1/2}$. Note that the matrix $\mathcal{Z}$ now has all real elements with a lower-dimensional representation (since $s < p$). In particular, each row of $\mathcal{Z}$ is a vector in $\mathbb{R}^s$. Thus the matrix $\mathcal{Z}$ represents a data cloud of p points in $\mathbb{R}^s$. We can label each row in $\mathcal{Z}$ as r so that the data cloud is given by $\{r_1, r_2, \ldots, r_p\}$. Given that the data cloud is defined in a proper metric space, we can partition the data cloud into k clusters, using method such as the k-means algorithm described late in this chapter (Section 4.5.3). A slightly modified form of the clustering technique can be found in Rohe et al. (2011).

In a sense, spectral clustering is not a clustering technique by itself. It is a way to convert similarity matrices into a geometric projection and then utilize regular clustering methodology to find the clusters. Figure 4.7 shows an example of such an application on Facebook connectivity across counties in the USA. See Guha et al. (2021) for an elaborate description of the algorithm for this application.

4.5.2 Hierarchies Based on Similarity

Hierarchical clustering generates tree-like structures of the data points/objects. Such structures are called *dendrograms* (Johnson, 1967; Rokach and Maimon,

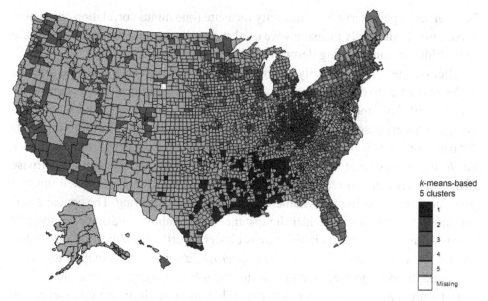

Figure 4.7 An application of *k*-means-based clustering on a spectral projection of county-to-county social media connectivity in the USA, 2016 (image created using the algorithm from Guha et al. [2021]).

2005). The main idea is to create a tree-like allocation of a given number of data points so that two points are put close to each other if they are similar with respect to some parameter. The tree structure borrows its name from tree graphs – networks that start with branches coming out from a single root and in turn branch out further, without creating any cycles. For the ease of visualization, typically these trees are drawn upside down so that the root node is at the top and the tree branches out below.

There are broadly two ways to carry out hierarchical clustering. One can either start at the entity level and assign clusters to entities, and then build up the structure by collecting similar entities into roots of small trees and connecting them further till the process iteratively connects all entities into a single tree. Such processes are called *agglomerative* clustering. The opposite method uses a *divisive* approach, where all entities are put in a single cluster first and then iteratively decomposed into smaller and smaller trees based on dissimilarity. The process continues until it exhausts all entities at the final *leaf* nodes.

Naturally, in order to carry out clustering, one first needs to define a similarity measure, or a dissimilarity measure for that matter. The dissimilarity measure between a pair of nodes can be defined in terms of any metric ranging from Euclidean to Manhattan to maximum distance among coordinates of the nodes. For non-quantitative features (e.g. dissimilarity among textual data), one can utilize metrics such as Hamming distance. For financial markets, return correlations between asset pairs serve as a similarity measure (one minus correlation coefficient serves as a dissimilarity measure – we used such a dissimilarity measure in the case of multidimensional scaling described earlier in the chapter).

After one gets a dissimilarity matrix, the next thing to decide is how to allocate nodes or entities to different clusters. To reiterate, for agglomerative clustering, we always start from individual nodes being clusters by themselves and recursively connect clusters based on minimum distance between clusters. The algorithms differ based on what is the definition of minimum distance. Under *farthest neighbor clustering*, the distance is defined as the maximum distance between all pairwise distances between two clusters. In this method, all nodes are connected under a given cluster, giving rise to its name *complete linkage* clustering. The opposite definition of distance is to consider the distance as the minimum distance among all pairwise distances between nodes across clusters. Further modification of the definition of distance comes in the form of *average* linkages, which exploit the value by taking the average over all pairwise distances between nodes across clusters.

In Figure 4.8, we present an example of hierarchical clustering on asset return data from the same set of stocks from the Bombay Stock Exchange that we analyzed in the case of multidimensional scaling. Mantegna (1999) was one of the pioneering analysts who proposed the application of hierarchical clustering to financial comovement networks. This technique has been widely used in subsequent literature on financial comovement networks.

4.5.3 Partitioning Data Clouds

Partitional clustering is a common and popular approach to finding a preassigned number of clusters in a data cloud. This technique requires the geometric coordinates of the data points to be predetermined and given. Crucially, this method differs from other clustering techniques in that the number of clusters needs to be specified *a priori*. The idea of this method is to consider data points to represent clusters to begin with and then iteratively decide whether the data points can be assigned to other clusters or not. The method minimizes a cost function based on the distances between points or between points and centroids of the emerging clusters. There are a number of variations of the key algorithm, going by the names of minimum k-clustering, k-center, and k-median, among others. Below we will define the key algorithm, which also turns out to be probably the most popular algorithm in this class. It goes by the name of k-means clustering (MacQueen, 1967).

Let us assume the dataset $X = \{x_1, x_2, \ldots, x_n\} \in \mathbb{R}^{p \times n}$ consists of n observations on p variables. Each x_i is a $p \times 1$ vector. The goal is to cluster these n data points into k groups where $k << n$. If k is close to n, then each point effectively becomes a cluster, which is not a particularly interesting case.

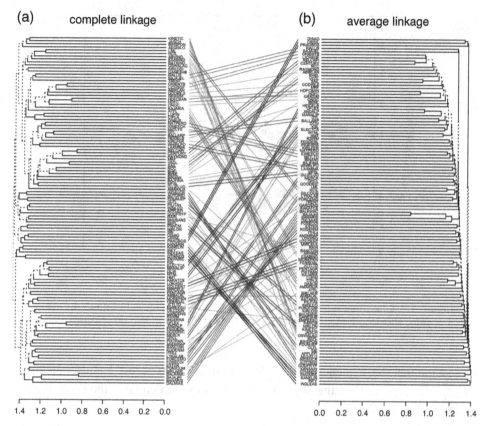

Figure 4.8 Comparative analysis of dendrograms: (a) complete linkage and (b) average linkage dendrogram of 115 stocks traded on the BSE (same set of stocks as in Figure 4.6). The lines represent the position of the same set of leaves present in both the trees.

We will give here a naive version of the algorithm to keep the presentation simple. There are many variations of this basic algorithm that we will not cover here (see Wagstaff et al., 2001; Kanungo et al., 2002; Likas et al., 2003). Let us assume that time is discrete and is indexed by t. The algorithm starts by randomly choosing k data points and assigns a label of *centroid* to them. Let us define the jth centroid at time $t = 0$ (i.e. the beginning of the algorithm) as $\mu_j^0 \in \mathbb{R}^p$. Then the algorithm calculates pairwise distances of all points from the centroids

$$d_{ij}^0 = ||x_i - \mu_j^0|| \tag{4.48}$$

for $i = 1, \ldots, n$ and then assigns each data point to a cluster centered around the centroid μ_j for which the pairwise distance is the minimum (i.e. the d_{ij}^0 is the lowest among all d_{il}^0 for $l = 1, \ldots, k$). Once all data points are assigned accordingly, let

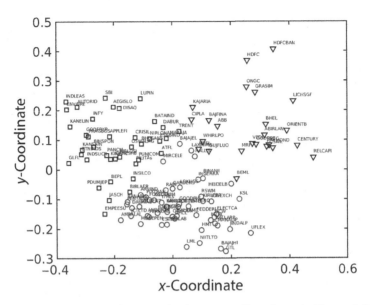

Figure 4.9 Classification of 115 stocks (same set of stocks as in Figure 4.6) into 3 groups, showing k-means clustering on the MDS map in two dimensions.

us assume that the sequence of the allocated data points across k clusters is given by $\{c_1^0, c_2^0, \ldots, c_k^0\}$. Then the algorithm proceeds by recalculating the centroids and updating the values:

$$\mu_j^1 = \frac{1}{|c_j^0|} \sum_{l \in c_j^0} x_l \qquad (4.49)$$

for $j = 1, \ldots, k$. Once we get the new set of centroids, the algorithm goes on to recalculate the distances and reassign data points based on the nearest distance. Thus the algorithm keeps on iterating from $t = 0$ to $t = 1$, and then to $t = 2$ and so on. Once the resulting allocation stabilizes (in the sense that the new allocation at time t remains the same as at time $t - 1$ subject to a pre-assigned tolerance level), then the algorithm ends.

Although the algorithm is reasonably fast and scalable to large datasets, it suffers from sensitivity to starting points and a lack of any guarantee of reaching global optima. An obvious follow-up question would be how to choose the parameter k? Pham et al. (2005) provide guidance on the choice of the parameter. A standard method is to use the elbow method of checking where the squared error (within the cluster) drops and becomes flatter. As an application, in Figure 4.9 we show the classification on the dataset that we analyzed in Figure 4.6.

4.5.4 Density-Based Clustering

Density-based spatial clustering of applications with noise (DBSCAN) is a density-based clustering technique (Ester et al., 1996). This method has two major advantages over partitional clustering (Tran et al., 2006; Zhou et al., 2000). First, this does not require a pre-specified number of clusters. Second, methods such as k-means do not deal well with spherical data clouds because they are dependent on distance from centroids by construction. Density-based clustering is relatively agnostic to the shape of the data and is more widely applicable to cluster data clouds with much more complex shapes.

Let us assume the dataset $X = \{x_1, x_2, \ldots, x_n\} \in \mathbb{R}^{p \times n}$ consists of n observations on p variables. Before describing the method, we need to define some terms. As is customary, we first define three types of data points with the help of two parameters: radius ϵ and a minimum number of points forming a cluster m. A data point x_c is called a *core* point if ϵ-neighborhood around x_c (given by $N_\epsilon(x_c)$) contains at least m number of data points in the data cloud. A data point x_b is called a *border* point if its ϵ-neighborhood contains less than m number of data points, and it is *reachable* from a core point x_c. Here, reachability from point x_i to x_j means that there exists a path from x_i to x_j such that each hop on the path is within ϵ-neighborhood of the previous data point. *Outliers* are data points x_o which are neither core nor border points.

The algorithm works as follows. It starts from a randomly chosen point x_r. If it has at least m points within its ϵ-neighborhood, then the first cluster is initiated at x_r. All the points within the ϵ-neighborhood of the points within the first cluster are also included in the first cluster, and so on. If the randomly chosen point x_r does not have m points within its ϵ-neighborhood, then it is temporarily tagged as an outlier (sometimes also called noise; note that the labeling in this step is temporary and can be revised later) and the algorithm proceeds to choose another point at random. Once the first cluster fully forms, then the algorithm randomly selects a previously unvisited data point and the same process begins, resulting in the second cluster and so forth. The algorithm stops when all points are part of some cluster or classified as outliers. We show an application in the Indian stock market in Figure 4.10 using the same dataset as in Figure 4.6.

4.6 Classifying Observations

Classification is a supervised machine learning approach. As the name suggests, the objective of the algorithms in this category is to classify entities based on some features.

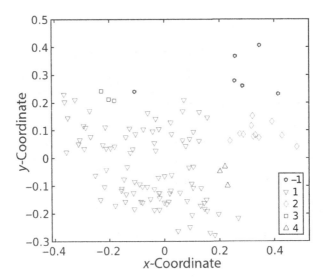

Figure 4.10 DBSCAN performed on a two-dimensional MDS map of 115 stocks traded on the BSE (same as in Figure 4.6) to cluster them based on a threshold. The threshold for a neighborhood search radius is $\epsilon = 0.08$ and the minimum number of neighbors $m = 3$ is required to identify the core point. Points described as -1 represent outliers.

4.6.1 Finding k-Nearest Neighbors

k-Nearest neighbors (often shortened as k-NN) is a classification algorithm. It is important to emphasize here that this is not a clustering algorithm and not directly related to k-means algorithm for clustering, although the names are quite similar. This algorithm is a very simple non-parametric method to classify new data points based on their relationship to nearby data points.

The algorithm works on a set of data given by $X = \{x_1, x_2, \ldots, x_n\} \in \mathbb{R}^{p \times n}$ which consists of n observations on p variables or features and a label vector given by c of size $m \times 1$ which maps each observation to a class. First, the algorithm requires an exogenously specified parameter k that will determine the number of data points required to qualify for a class. Given a new data point x_{n+1}, the algorithm first calculates all pairwise distance $||x_i - x_{n+1}||$ for $i = 1, 2, \ldots, n$ (typically a Euclidean norm is used for continuous variables; for other types of variables such as textual data, Hamming distance can be used) and chooses k data points that have the least distance from x_{n+1}. Within these k data points, the idea is to calculate the relative frequency of data points belonging to different classes. Under the *majority rule*, whichever class dominates within those k-neighbors, the algorithm assigns the new data point to that class.

There is no *a priori* way to choose k. The parameter is context-dependent. Generally, smaller values of k lead to higher variance whereas larger values of

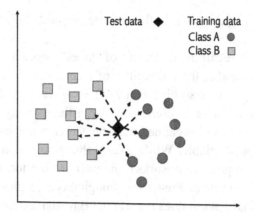

Figure 4.11 Schematic representation of *k*-nearest neighbors. Two types of data are classified as class A (shown as a circle) and class B (square). A new data point (diamond) will be classified as part of either A or B based on the Euclidean distance among its *k* nearest neighbors.

k lead to higher bias in classification. Often it is data that drives the choice of *k*. Sometimes heuristics along with cross-validation methods are used to find *k*. One obvious criterion for choosing *k* is to choose an odd number in order to avoid ties.

We show a schematic of the method in Figure 4.11 where the data in the scatterplot is categorized in two different classes. Now, a new data point with unknown class type is added to the existing data. By using the *k*-NN method, we can predict the class of the new data point based on its neighbors. For very high-dimensional data (say *p* is in the order of 10), this method does not perform well since all points get scattered almost equally far away from the new data point. This method works much better in smaller dimensions.

4.6.2 Classification Using Bayes' Theorem

Naive Bayes classifiers are a group of classification algorithms that work by utilizing Bayes' theorem (Hastie et al., 2009; John and Langley, 1995; Rish, 2001). We have already encountered this theorem in our discussion of statistics (see Chapter 2). Here we will see that the same theorem provides a powerful tool for classification as well.

The main idea behind the algorithm is to find the conditional likelihood of classification in a certain group based on a given set of features, where the training dataset provides mapping between the features and the resulting classes. Let us denote these features by a vector $\mathbf{f} = \{f_1, f_2, \ldots, f_n\}$ and the groups or classes by another vector $\mathbf{g} = \{g_1, g_2, \ldots, g_k\}$. The goal is to figure out what is the conditional probability of an entity being assigned to a given group g_j based on observed features $\{\hat{f}_1, \hat{f}_2, \ldots, \hat{f}_n\}$. In the language of probability, the goal is to find $p(g_j|\hat{f}_1, \hat{f}_2, \ldots, \hat{f}_n)$.

Here it is useful to recall the statement of Bayes' theorem. The theorem states that for two events x and y, the probability of x conditional on y can be found as $p(x|y) = p(y|x)p(x)/p(y)$ assuming that $p(y) \neq 0$ (refer to Equation 2.4). The Bayes classifier exploits this theorem to evaluate the quantity $p(g_j|\hat{f}_1, \hat{f}_2, \ldots, \hat{f}_n)$. It is called *naive* because it assumes that the features are independent of each other, given the group variable. While clearly this assumption might be violated for real-world data, in practice it works quite well. Before formally describing the derivation, we note two things. First, even though this algorithm falls into the category of Bayesian classifiers, it does not use the Bayesian concept of probability or even estimation. Regular parametric assumptions lead to maximum likelihood estimators. Second, the independence assumption (which makes the algorithm naive) reduces the computational cost significantly, as we will show below. But it may not hold in reality.

Let us start with some known set of features $\{f_1, f_2, \ldots, f_n\}$ and known set of groups $\{g_1, g_2, \ldots, g_k\}$. There is a training dataset that matches the features to the groups. Based on the matching, the algorithm has to classify an entity into one of the groups within the given k number of groups, based on a vector of observed features $\hat{\mathbf{f}} = \{\hat{f}_1, \ldots, \hat{f}_n\}$.

We can start by applying Bayes' theorem, obtaining the conditional probability that the entity belongs to the jth group as

$$p(g_j \mid \hat{\mathbf{f}}) = \frac{p(\hat{\mathbf{f}} \mid g_j)p(g_j)}{p(\hat{\mathbf{f}})}$$

$$= \frac{p(\hat{f}_1, \ldots, \hat{f}_n \mid g_j)p(g_j)}{p(\hat{f}_1, \ldots, \hat{f}_n)}. \tag{4.50}$$

The key thing to note here is that the term in the numerator $p(\hat{f}_1, \ldots, \hat{f}_n \mid g_j)$ can be simplified to the following (using conditional independence of features):

$$p(\hat{f}_1, \ldots, \hat{f}_n \mid g_j) = p(\hat{f}_1 \mid \hat{f}_2, \ldots, \hat{f}_n, g_j)p(\hat{f}_2 \mid \hat{f}_3, \ldots, \hat{f}_n, g_j) \cdots$$
$$\cdots p(\hat{f}_{n-1} \mid \hat{f}_n, g_j)p(\hat{f}_n \mid g_j)$$

$$= \prod_{i=1}^{n} p(\hat{f}_i \mid g_j). \tag{4.51}$$

The last step utilizes the independence assumption to write the conditional probabilities as

$$p(\hat{f}_i \mid \hat{f}_{i+1}, \ldots, \hat{f}_n, g_j) = p(\hat{f}_i \mid g_j) \tag{4.52}$$

for $i = 1, 2, \ldots, n$. This assumption is fairly strong and may not be realistic. But as one can see, this makes the calculations substantially easier. Under the above assumption, Equation 4.50 can be rewritten as

$$p(g_j \mid \hat{\mathbf{f}}) = \frac{p(g_j) \prod_{i=1}^{n} p(\hat{f_i} \mid g_j)}{p(\hat{\mathbf{f}})}. \tag{4.53}$$

Note that the denominator is independent of the group or class assignment and therefore it remains the same across groups.

Now, we can evaluate the probability of an entity belonging to a certain group or class given a vector of observed characteristics, where each component of the probability can be calculated from the observed mapping, that is, from the training dataset. Two things are still unresolved. The first one is how to actually evaluate the probability: in other words, how do we assign numbers to the expression given in Equation 4.53? The second question is: how do we convert the probability into a decision on whether or not an object belongs to a given class or not? For the first question, one can make parametric assumptions. For example, one can assume that the probability distributions are normal. For the second, one can simply estimate the probability for all classes given the observed characteristics $\hat{f}$, and assign the class which has the highest conditional likelihood.

4.6.3 Using Support Vectors

The support vector machine (often shortened as SVM) is a supervised classification algorithm (Boser et al., 1992; Noble, 2006). It has gained widespread popularity due to its high accuracy and ability to deal with high-dimensional data (Vert et al., 2004; Schölkopf et al., 2002; Shawe-Taylor and Cristianini, 2004).

The goal of the algorithm is to classify data points into more than one class using a geometric concept of separation. Let us denote the dataset by $X = \{x_1, x_2, \ldots, x_n\} \in \mathbb{R}^{p \times n}$ which consists of n observations on p variables. Each data point comes with a label indicating the class it belongs to. Let us assume that there are two classes denoted by $+1$ and -1, which is customary in this literature. We will follow this convention.

Hyperplanes and Linear Classifiers

Let us consider a simple case of a linear two-class classifier. Let us also imagine that the data cloud is clearly separated geometrically. One way to explore the separation is to consider a *hyperplane* passing through the cloud so that the two classes of data points fall on two different sides of the hyperplane. The equation of the hyperplane is written as

$$\phi'x + b = 0 \tag{4.54}$$

where ϕ is a vector and b is a scalar. Clearly, if $b = 0$ in Equation 4.54, the hyperplane goes through the origin. It may be checked that the vector ϕ is orthogonal to the hyperplane. For an intuitive picture, one may note that a hyperplane in two dimensions is a straight line and a hyperplane in three dimensions is a two-dimensional plane. A general linear discriminant function (that allows us to discriminate between the classes) is written as

$$f(x) = \phi'x + b. \tag{4.55}$$

The vector ϕ is called the *weight vector*, and b is called the *bias*. The magnitude of the bias b determines the distance of the hyperplane from the origin.

Note that the set of points on the hyperplane

$$S = \{x : f(x) = \phi'x + b = 0\} \tag{4.56}$$

splits the space into two subspaces. For any given point x, the sign of $f(x)$ denotes the side of the hyperplane the point would be on. Now it is useful to introduce the concept of the *decision boundary*. This is effectively the separator that allows us to classify the regions as positive and negative for the purpose of the algorithm. Clearly, a separating hyperplane is a good candidate for decision boundaries. Note that there can be more than one separating hyperplane and therefore such separators are not necessarily unique. We show an example of a linear classifier in Figure 4.12. Further generalizations to nonlinear classifiers is also possible.

Before proceeding further, we first define the norm of the weight vector ϕ as $||\phi|| = \sqrt{\phi'\phi}$. A unit vector in the direction of ϕ would be given by $\hat{\phi} = \phi/||\phi||$. Note that the distance of the hyperplane from the origin is given by $b/||\phi||$. Now, let us denote the two points belonging to the two classes that are closest to the hyperplane as x_+ and x_-. The best classifier should ideally split the distance equally, resulting in two equations $\phi'x + b = a$ and $\phi'x + b = -a$ for some scalar a. Now, note that the distance between the two points is the projection of $x_+ - x_-$ in the direction of $\hat{\phi}$, which results in the distance being written as

$$d_{x_+,x_-} = (x_+ - x_-)\hat{\phi}. \tag{4.57}$$

By substituting for $\hat{\phi}$, we see (ignoring the constant) that

$$d_{x_+,x_-} \propto \frac{1}{||\phi||}. \tag{4.58}$$

Therefore, to maximize distance, we need to minimize $||\phi||$.

However, this is not an unconstrained optimization problem. In order to correctly separate the classes, we need class $c_i = +1$ to be assigned to all points x_i for which $(\phi'x_i + b) \geq a$ and $c_i = -1$ to be assigned to all points x_i for which $(\phi'x_i + b) \leq a$.

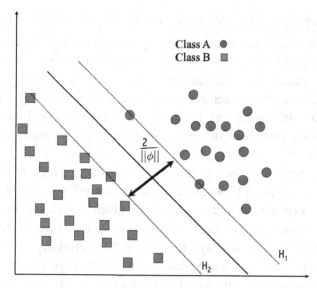

Figure 4.12 Schematic representation of support vector machine, using a linear classification of two types of data as class A (circle) and class B (square). The equation of the hyperplane is given by $\phi'x + b = 0$. The solid line shows a clean separation. The dotted lines closest to the two data clouds are the margin lines.

The quantity a can be normalized to 1. Then the combined constraint can be written as

$$c_i(\phi'x_i + b) \geq 1 \tag{4.59}$$

for $i = 1, 2, \ldots, n$. Therefore, one can write the constrained optimization problem as (see e.g. Ben-Hur and Weston, 2010)

$$\begin{aligned} \min_{\phi, b} \quad & \|\phi\| \\ \text{s.t.} \quad & c_i(\phi'x_i + b) \geq 1 \quad \text{for } i = 1, \ldots, n. \end{aligned} \tag{4.60}$$

Let us denote the estimated ϕ and b by adding a hat sign. Then given a new data point x_{n+1}, the sign of the expression $\hat{\phi}'x_{n+1} + \hat{b}$ determines the classification.

In the above discussion, we have assumed linear separability throughout. In many real-world applications, the separation may not be found linearly. The literature has developed many more modified algorithms based on the above principle to tackle such scenarios. A standard approach for nonlinear separability is to project the given dataset into higher dimensions by augmenting the space of the data cloud, such that the classes become linearly separable in the higher dimensional space. Interested readers can find more detailed expositions and applications in Suykens and Vandewalle (1999), Amari and Wu (1999), Tong and Koller (2001), Tong and Chang (2001), and Fung and Mangasarian (2005).

Separating Periods of High and Low Volatility in Financial Markets

Volatility is an emergent feature of agents interacting and trading in financial markets. High market volatility is generally thought to represent times of turbulence and low volatility to represent calm periods. Here we apply a support vector machine to differentiate between periods of high and low volatility. This is a somewhat unconventional application. Typically, support vector machines are applied to image recognition, voice recognition, and so on. But for our purposes, we will stick to this example.

We have collected data on the Indian stock market at high frequency for 50 stocks traded in the NIFTY 50 index in 2020. We will not describe the market here. A longer explanation and description is given in Chapter 3 around Figure 3.4. For our purposes, it suffices to note that the data is in panel format – for 50 stocks we have time series data synchronized at 10-s intervals. We have such data for 30 consecutive trading days. We have excluded the days when the market was closed for trading, typically for national holidays and weekends. For each day, we split the data into three trading periods – morning (9 a.m. to 11:30 a.m.), afternoon (11:30 a.m. to 1:30 p.m.), and evening (1:30 p.m. to 3:30 p.m.). For each of these time zones, we compute two measures of volatility. First, we compute the standard deviation for each asset's return and take the average across 50 stocks. Second, we compute the standard deviation of centrality estimated from a network created using a distance matrix. We will not discuss the construction of the network here; we described it in the discussion around Figure 3.4. We also omit discussion of centrality here but will cover it in Chapter 5.

For each trading period, we get two values – average return dispersion and centrality dispersion on the network. For our purpose here, all we need to consider is that these are two variables denoting two different forms of dispersion in the market, one directly capturing fluctuations while the other captures dispersion in participation of different assets in driving aggregate market dynamics.

Note that for each day, we get three data points. Therefore, for 30 days, we get 90 data points in total. Now we classify the periods of high and low volatility based on the product of the two measures, such that if the product is less than the *median* of that variable then it's marked as a period of low volatility; otherwise we call it a period of high volatility. The classification is purely hypothetical and can be altered. However, it is useful in thinking about dispersion of volatility in financial markets. When markets open and close, they are often more volatile than during the day. Thus a cut-off can potentially segregate these periods nicely. Given this criterion, we can now classify the periods into two regimes of high and low volatility.

In Figure 4.13, we show the scatterplot on a log–log graph. We have taken the log to convert the product of volatilities (used for classification) into a linear criterion.

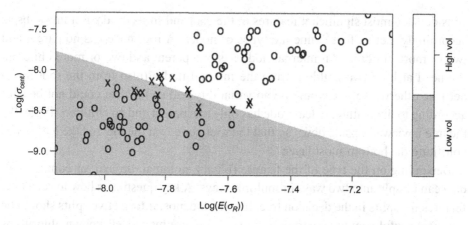

Figure 4.13 Using a support vector machine for separating market volatility regimes. The gray region represents low volatility and the white region represents high volatility. Circles and crosses represent different stocks at different trading times. We analyzed the same set of stocks from a network perspective in Figure 3.4.

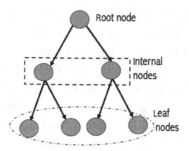

Figure 4.14 Schematic representation of decision tree at three levels: root node, internal nodes, and leaf nodes.

Now, suppose one does not know the criterion for classifying a given data point in the regime of high or low volatility. In that case we can utilize a separator between them so that a new data point can be classified by following the separator. We superimpose the separator implied by the support vector machine on the same plot. This is estimated with a linear kernel in the R programming environment.

4.6.4 Decision Trees

A decision tree is a type of supervised learning algorithm that exploits the features (also known as attributes) of a given dataset in a recursive manner (Alpaydin, 2020). As the name indicates, decision trees are constructed as a tree-like structure with a root, internal nodes, and leaf nodes (Figure 4.14). The algorithm recursively considers the most significant features of the data and splits the dataset into subsets accordingly. Let us first define the types of nodes. A root node has no parent and two or more children. An internal node has one parent and two or more children. We need at least two children from the internal nodes (also from the root node) because otherwise there would be no variability and so the data could not be split according to the feature. A leaf node has only one parent and no children. Typically, the tree is viewed upside down so that the root node is at the top and the leaf nodes constitute the bottom-most layer.

Depending on the type of the feature (binary, nominal, ordinal, or continuous), data can be split into two ways or multiple ways. A key question is how to select the features for splits in the decision tree. Ideally, the most informative splits should be the most useful way to construct the tree. There are many well-known algorithms such as Hunt's algorithm, iterative dichotomiser 3, C4.5, and the classification and regression tree (often shortened as CART) (Quinlan, 2014). Below, we consider a simple case and describe a decision tree for entities that come with a given number of features and where the outcomes are discrete.

Let us consider a sample S which can be split according to some feature f into c classes. Let p_i denote the fraction of entities belonging to class i for $i = 1, 2, \ldots, c$. We need to define a degree of impurity for such splits, which indicates the level of determination (actually, lack thereof). Intuitively, the smaller the degree of impurity, the more skewed would be the class distribution. For example, a node with distribution 0 and 1 across two classes has zero impurity. The converse scenario, with distribution 0.5 and 0.5 across two classes, has the highest impurity. An index for impurity is given by entropy at a node s:

$$Ent_s = -\sum_{i=1}^{c} p_i \log_2 p_i. \tag{4.61}$$

One can define two more commonly used impurity measures. *Gini impurity* is used by the CART algorithm for classification trees and is given by

$$G_s = 1 - \sum_{i=1}^{c} p_i^2, \tag{4.62}$$

which is bounded between 0 and 1. Equal distribution (implying high impurity) maximizes G_s whereas completely unequal distribution (only one class gains all mass and the rest get zero, implying low impurity) has the least G_s (in which case, the value equals 0). The other useful metric is *information gain* which is based on the decrease in entropy after a dataset is split according to a feature and is defined as

$$I_s = Ent_s - \left(\sum_{i=1}^{c} \frac{n_i}{n} Ent_i \right) \tag{4.63}$$

where a node s is split into c partitions and n_i is the number of entities in class i.

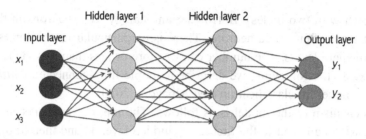

Figure 4.15 Schematic representation of a neural network, consisting of an input layer, two hidden layers, and an output layer. In a fully connected network, all of the nodes in consecutive layers will be connected by directed arrows. For clarity, we show fewer arrows here. In more complex models, the arrows indicating the flow can go backward as well.

The split depends on which feature leads to the largest reduction in impurity (information gain), or whichever feature leads to the lowest impurity (Gini). Information gain is used extensively in decision tree algorithms, which tend to prefer splits that result in large number of small partitions which have low impurities. Finally, one can define classification error as $1 - \max_j p_j$, which gives a very easy way to calculate how precise the classifications are.

Here we omit a full-fledged discussion on this topic as these kinds of classifiers have not been used a lot in the context of complex systems to the best of our knowledge. Readers interested in an in-depth review of the technique may consult Tan et al. (2006) and Alpaydin (2020).

4.6.5 Artificial Neural Networks

Artificial neural networks represent probably the most well-known algorithm in the toolbox of machine learning; the idea has gone beyond the confines of academia and captured public imagination. In part, the attraction is due to the seemingly rich analogy with the human brain, at least on a superficial level. There has been enormous progress in the study of learning models, thanks to artificial neural networks. The subject is so vast that it is impossible to summarize its whole development here. The applications of neural network models are incredibly rich in variety and they can work as both supervised and unsupervised learning environments. Instead of trying to describe the general class of models, we will pick up a toy model for classification purposes using a backpropagation mechanism, and we will explain the key mechanism below. Going forward, we will drop the word "artificial" and describe these models simply as neural networks.

Before describing how the model works, it is useful to describe its underlying architecture. In Figure 4.15, we have plotted a schematic representation of a neural network. It consists of an *input* layer (shown with three nodes), two intermediate layers which are described as *hidden* (shown with four nodes each), and an output layer (shown with two nodes). Borrowing an analogy from neurons in the brain, these nodes are often called neurons. There is no particular reason for us to have two intermediate layers. The number can be more or less than that, depending on the context. In the following, we describe a simple fully-connected *multi-layered perceptron* model for classification (see Alpaydin, 2020).

The mechanism of the model is as follows. The input layer represents a set of impulses, which are sent to the first layer (hidden layer 1) and then to the second layer (hidden layer 2). Each of the nodes in these layers transforms the impulses (we will describe the transformation later) and sends the transformed impulses to the downstream layers. Finally, the transformed impulses reach the output layer. The directions of flows of impulses are shown by the arrows. It is useful to mention here that the number of hidden layers or even the number of nodes in the layers is often context-dependent. The input layer has the same number of nodes as the number of different impulses. For a classification task, the output layer typically has the same number of nodes as the number of classes within which the impulses should be classified.

Now, we are in a position to describe the transformation and the *training* phase. Following the figure, we start with three impulses represented by three nodes in the input layers, given by $\{x_i\}_n$ ($n = 3$ in the figure). For any given node z at any hidden layer, let us denote the incoming vector of impulses by x. The transformation is given by the *activation function*:

$$f(\xi) = \frac{1}{1 + exp(-\theta\xi)} \tag{4.64}$$

where the variable ξ is the sum total of all impulses coming to node z:

$$\xi = w'x + b \tag{4.65}$$

where w is a vector of weights attached to the paths of impulses, and b is a constant. An alternate functional form is also quite common in the form of a hyperbolic function

$$f(\xi) = \frac{1 - exp(-\theta\xi)}{1 + exp(-\theta\xi)}. \tag{4.66}$$

Both functions are sigmoidal in nature. The parameter θ determines the gradient of the activation function. Another standard activation function that is often used is $f(\xi) = \max\{0, \xi\}$. With the above formulation, the vector of the triplet of parameters $\{\theta, w, b\}$ contains free parameters that can be tuned, which makes the model *learn* about the relationship between the input and the target output. We will now describe how the tuning is done.

Each node in each layer can be endowed with an activation function. Thus the initial set of impulses will get transformed multiple times before appearing as the output through the output layer. Let us denote the output by a vector $\{y_j\}_m$ following the figure ($m = 2$ in the figure). During the training phase, correct outputs are specified. Let us denote them by $\{c_j\}_m$. The goal is to minimize the difference between the produced output and the correct output by tuning the free parameters.

For that purpose, let us define an error term capturing the gap between the predicted output and the true output:

$$e_j = c_j - y_j, \tag{4.67}$$

so that the total error at the final layer (let us denote a layer by l) is given by $e^l = \sum_j e_j^2$. One can define aggregate error across all layers as $e = \sum_l e^l$. Once the errors are calculated, they are *backpropagated* (hence the name of the algorithm) at every layer and the weights are adjusted at the node level to reduce errors. A standard technique is to use numerical adjustments on the parameter vector. The adjustment mechanism works as follows. Let us denote a generic parameter by p. Then we evaluate reduction in error by adjusting the parameter by a small amount

$$\Delta p = -\tau \frac{\delta e}{\delta p} \tag{4.68}$$

and tune the parameter by updating it as

$$p_{new} = p_{old} + \Delta p. \tag{4.69}$$

The new parameter τ captures the *learning rate* of the algorithm: that is, it determines how quickly parameters can adjust in order to reduce the error.

The algorithm is initiated with some arbitrary parameter configuration (the initial choices can also be informed in some case if more information is available) and then the algorithm goes back and forth following the above steps until the model has *learned* to classify correctly (subject to some exogenously specified criteria) based on an input vector. Once the model is ready, then it can be used on test data.

The above is a rudimentary discussion and far from complete. The literature on and around neural networks has seen explosive growth. For further discussion, readers can refer to Haykin (1994). In many books on machine learning, neural networks feature very prominently. There are too many references to list here. We would like to point the interested reader to Russell and Norvig (2002) which has become a classic in this literature and presents a very useful treatment of neural networks in the context of artificial intelligence.

Finally we note a point here about the success of this class of learning models. Neural networks have been tremendously successful in pattern recognition. Given that large networks function like a black box, a question that naturally comes up is exactly why such networks are so good at finding and learning patterns. A fundamental building block of the answer is the *universal approximation theorem* which loosely translates into the idea that any nonlinear function can be approximated well by a sufficiently large number of activation functions with varying weights. The literature around this question is very active and newer developments are taking place at a rapid pace toward a more complete understanding. We end this discussion with a reference to Moravec's paradox that tasks based on reasoning are computationally tractable whereas seemingly simpler tasks like perception have proved to be much more challenging for computational feasibility.

Sentiment as a Latent Variable of Complex Interactions

Complex interactions lead to emergent outcomes. Often, for economic and social systems, such outcomes depend on the underlying motivation and behavior of the agents. One strand of the literature has attempted to quantify this phenomenon in the form of sentiment analysis. Sentiment can be thought of as representative of the polarity of a news item. It may itself arise out of agent-level interactions or it may be exogenous and lead to agent-level interactions. Either way, it can be treated as a latent variable that is not directly observed.

As an application of the classification algorithms, we consider sentiment analysis on financial texts. We use the data produced by the algorithm given in Malo et al. (2014) (see also Sinha et al. [2022] and references in both of these two papers) for the classification of financial texts according to their expressed *sentiment*. Specifically, the underlying data is a sequence of news headlines that appeared in digital media. The data is aggregated at the level of a minute and then classified as positive $(+1)$ or negative (-1). Neutral news is classified as 0, and the minutes when a news item does not appear are also classified as 0.

In Figure 4.16, we show the evolution of sentiment as interpreted by the classifier in its level (top panel), averaged over time (middle panel) and cumulative over time (in the bottom panel) in a given day. The timing is from 12:00 midnight to 11:59 p.m. From midnight until morning, there are not many news items. In the morning news items start appearing, and during the day the sentiment shows a substantial amount of fluctuation. The x-axis starts from a non-zero number because this figure is a snapshot: we have ignored the previous days in this plot.

In Figure 4.17 we show evolution of daily sentiment over a year and how it relates to the corresponding financial market. Specifically, we show that the daily sensex return (index for the Bombay stock market) changes with respect to changes in the daily sentiment score. We have not shown the estimates here, but a linear regression shows that the relationship is significant.

4.7 Model Validation and Performance

Machine learning techniques are useful for highly nonlinear problems that do not possess analytical solutions. Also, in many cases, the nature of the global solution may not be *a priori* clear. Since the techniques we have discussed above

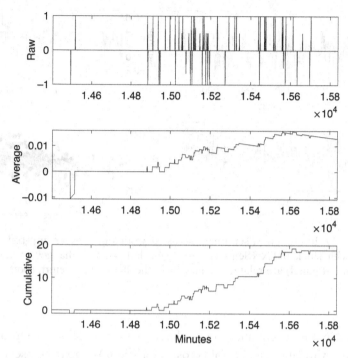

Figure 4.16 Evolution of sentiment in a given day across minutes in the Indian market as captured by analyzing sequential news headlines appearing in the online media. Raw (top panel) indicates the level of sentiment categorized as positive (+1), negative (−1), or neutral (0). If no news came in during a given minute, then we take the sentiment to be 0. Average (middle panel) indicates sentiment value averaged from the starting minute. Cumulative (lowest panel) indicates cumulative sentiment from the starting minute.

are predominantly numerical in nature, there is often no theoretical guarantee of reaching global optima. Additionally, since data trains the model, there is a substantial chance of having too large in-sample fit resulting in bad out-of-sample performance. In such cases, it is useful to have access to methods to evaluate the robustness of the models' predictions. Resampling methods are techniques that allow the modeler to repeatedly sample data and serially evaluate the models in order to potentially make the predictions more robust.

Cross-Validation

Once we fit a model to a given dataset, we often want to know how robust the parameter estimates are. Or we may ask whether the fitted model is the best among the available models. Cross-validation is a useful technique for addressing such queries (Kurtz, 1948; Mosier, 1951; Krus and Fuller, 1982). There are three easily

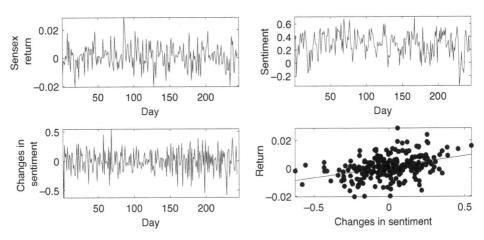

Figure 4.17 Evolution of sentiment across days in a given year (top-right panel) along with stock index (Sensex) return (top-left panel). Changes in sentiment (bottom-left panel) are related to changes in the stock index return (bottom-right panel).

implementable ways to conduct cross-validation: the validation set approach, leave-one-out cross-validation, and k-fold cross-validation. We give the pseudocodes for the first two approaches below. k-fold cross-validation is a generalization of the leave-one-out cross-validation. Additionally, we note that bootstrapping is another cross-validation technique. This is fundamentally a statistical resampling technique based on sampling with replacement (Efron, 1979, 1981, 1982). In the present discussion, we omit it as it falls outside the scope of this chapter. Interested readers can consult Tibshirani and Efron (1993).

Algorithm 4.1 *The validation set approach*

---

1. Partition the dataset randomly into a training set and a validation set.
2. Fit the model on the training set.
3. Apply the fitted model to the validation set to make predictions.
4. The mean squared error of the predictions provides an estimate of the error rate.

Ensemble of Models

Ensemble learning is a particular type of machine learning paradigm that exploits the power of multiple models with the goal of solving a given problem (Hansen and Salamon, 1990; Schapire, 1990; Zhou, 2012; Kuncheva, 2014). Ensembles may work by a sequential arrangement or a parallel arrangement between the learning models. When each of the learning models makes predictions, there has to

---

Algorithm 4.2 *Leave-one-out cross-validation*

---

1. Construct the validation set from a single observation (x_1, y_1).
2. Construct the training set from the remaining observations $\{(x_2, y_2), \ldots, (x_n, y_n)\}$.
3. The model under consideration is fitted on the $n-1$ observations in the training set, and a prediction $\hat{y}_1$ is made x_1 using the fitted model.
4. The mean squared error $\text{MSE}_1 = (y_1 - \hat{y}_1)^2$ provides an estimate for the error.
5. Repeat steps (1 through 4) by excluding the residual data points (x_i, y_i) and store the corresponding MSE_i for $i = 2, \ldots, n$.
6. The resulting MSE is given by

$$LCV = \frac{1}{n} \sum_{i=1}^{n} \text{MSE}_i. \tag{4.70}$$

---

be a decision rule on aggregating potentially different predictions into a single number of classes. For classification purpose, typically the *max voting* rule works efficiently. In this method, multiple models are used to make predictions and each prediction is considered to be a *vote*. The decision rule picks the prediction that gets the majority of the votes as the final prediction. For continuous data (e.g. regression) a *weighted average* works well, where the idea is to assign weights to different predictions arising out of different models, taking the average to come up with an aggregate prediction (Alpaydin, 2020).

Popular ensemble-based methods are bagging (Breiman, 1996), boosting (Schapire, 1990), stacking (Wolpert, 1992), and AdaBoost (Freund and Schapire, 1996), among others. Several popular extensions have been quite successful as learning models (e.g. random forest). The main insight used in ensemble learning is that often individual models do not perform very well due to the bias–variance trade-off (recall Figure 4.1), but a collection of models performs substantially better. Bagging, boosting, and stacking are three different ways of finding *ensembles* of different models to produce superior performance. Bagging considers homogeneous models to train them on different subsets of data, and combines the resulting fitted models to produce an aggregate model. Fundamentally, bagging is a parallel learning technique. Boosting, on the other hand, considers homogeneous models to train them sequentially on different subsets of data, updating the fitted models while switching from one training dataset to another. Finally, stacking combines different models in parallel learning from different subsets of data. For more in-depth discussion, interested readers may consult Opitz and Maclin (1999), Polikar (2006), Bishop (2006), Rokach (2010), and Jordan and Mitchell (2015).

Table 4.1 *Confusion matrix: actual values are either positive or negative, and predicted values are either positive or negative.*

		Actual	
		Positive	Negative
Predicted	Positive	TrPos	FaPos
	Negative	FaNeg	TrNeg

Evaluating Performance

Any model that makes predictions suffers from the possibility of making wrong predictions – there can be false positives or false negatives. Similar to Type I and Type II errors in statistics (see Section 2.3.2), classification algorithms also come with misclassifications of both types.

A standard way to capture the misclassifications is via the *confusion matrix* or the *error matrix*. Table 4.1 represents the confusion matrix, where rows correspond to predicted values and columns correspond to actual values. TrPos stands for positive positives (correctly identified positive state), FaNeg stands for false negatives (incorrectly identified positive state), FaPos stands for false positives (incorrectly identified negative state), and TrNeg stands for true negatives (correctly identified negative state).

The *Accuracy* of a model is given by the number of correct predictions divided by the total number of predictions:

$$Acc = \frac{TrPos + TrNeg}{TrPos + FaPos + TrNeg + FaNeg}. \tag{4.71}$$

The *Precision* of a model is defined by how frequently the model makes correct predictions about true states when it makes a positive identification:

$$Prec = \frac{TrPos}{TrPos + FaPos}. \tag{4.72}$$

Recall or *sensitivity* measures the proportion of true states that are correctly identified as positives:

$$Rec = \frac{TrPos}{TrPos + FaNeg}. \tag{4.73}$$

The *F-score* is a combination of precision and recall:

$$F\text{-}score = \frac{2 \times Prec \times Rec}{Prec + Rec}. \tag{4.74}$$

Specificity measures the proportion of false states that are correctly identified as negatives:

$$Spec = \frac{TrNeg}{FaPos + TrNeg}. \tag{4.75}$$

We can also utilize graphical tools for diagnostic purposes by combining the information contained in the measures defined above. Two well-known measures are the ROC curve and the corresponding AUC values. The receiver operating characteristic (ROC) curve is a plot with two axes, which shows the performance of a binary classifier as a function of its cut-off threshold. It captures the true positive (TrPos) rate (given by *Rec* in Equation 4.73) against the false positive (faPos) rate (given by 1–*Spec* from Equation 4.75) for varying discrimination threshold values. Both the axes have a scale of (0, 0) to (1, 1). The area under the curve (AUC) is a measure of performance of a binary classifier on all possible threshold values. This is found by integrating the area under the ROC curve.

4.8 Taking Stock and Further Reading

Machine learning has been remarkably useful across many disciplines to find patterns in large-scale and messy datasets (Bishop, 2006; Alpaydin, 2020). In this chapter we have described a number of standard machine learning techniques ranging from clustering to spectral analysis to nonlinear learning algorithms. There are many excellent textbooks on machine learning. James et al. (2013) is a very useful reference on machine learning from a statistical perspective. For a larger perspective on the artificial-intelligence-oriented approach, readers can consult Russell and Norvig (2002) which is a classic in this literature. One of the biggest omissions from our review of machine learning topics is NLP or natural language processing. We have deliberately omitted the topic as NLP does not yet have frequent and direct applications to the study of complex systems. Having said that, it is quite possible to see newer applications coming up in the context of language dynamics (see the model reviewed in Chapter 6) in the near future. Silva and Zhao (2016) present some recent developments in machine learning in the context of complex networks.

We have completely avoided discussing reinforcement learning. This paradigm is an extremely important component complementing supervised and unsupervised learning. Loosely speaking, the idea is to utilize a payoff function for a learner who tries out different policies to optimize an objective function: the model reinforces a policy choice which generates higher payoffs while retaining scope for further explorations and tinkering with the policy. Therefore, the eventual solution emerges through a balance between exploration and exploitation. Sutton and Barto (2018) is an excellent resource for this learning paradigm. At a theoretical level, one classic problem called the *bandit problem* captures the idea quite nicely. The problem is to allocate resources over multiple choices in search of higher expected efficiency. The catch is that the properties of the choices become known only after allocations are made: that is, exploration occurs, which may reduce efficiency. The famous Gittins Index (Gittins, 1979) provided an early solution to this problem, and a large literature developed around it.

More generally, the current literature has unearthed many new dimensions including optimal stopping problems, dynamic programming, and game theory. Jordan and Mitchell (2015) provide a summary of the trends and new directions in the machine learning literature. A new direction important for complex systems has been uncovered in the recent literature on using machine learning for statistical inference. Machine learning by itself is geared toward predictive modeling. Specifically, as long as a technique allows us to predict accurately, the algorithm is thought of as useful. However, in many cases, the exact nature of the relationship between the input variable and the output variable is more important. This is especially true for establishing causality. Unfortunately, such statistical inference does not generally coincide with prediction. This problem is often known as the $\hat{\beta}$ (inference) versus $\hat{y}$ (prediction) problem. There is a large literature that attempts to bridge the gap between machine learning, statistics, and econometrics by directly addressing the dual issue of inference and predictions. Interested readers may consult Quinlan (2014), Mohri et al. (2018), Mitchell (1997), Sammut and Webb (2011), and Murphy (2012). Finally, we note that machine learning is becoming quite mainstream in social sciences, especially in economics. See Mullainathan and Spiess (2017) and Athey and Imbens (2019) for two relatively recent reviews.

5

Interlinkages and Heterogeneity
Network Theory

5.1 Understanding Linkages

One of the key signatures of a complex system is the emergence of patterns which cannot be understood by isolating the constituent parts of the system. It is the interaction among elements of the system that leads to the emergence of patterns. Network theory has become an integral part of the study of complex systems due to the simplicity of describing connections of a very complicated nature through a standardized mathematical object. At the same time, network theory has made its way into the data science literature due to its algebraic simplicity and manipulability. As we will show, matrix algebra can take us far into studying the network structure of a large variety of interlinkages that can be found in the physical, social, and economic worlds. The widespread applicability of network theory along with its ability to compress information has made network theory a suitable candidate for tying data science together with complex systems. In what follows, we will first provide an intuitive picture of exactly what networks are and then we will proceed toward describing the methodology behind network theory.

In an abstract sense, a network is a collection of a set of nodes and the edges connecting them. The definition is quite general, allowing us to express a wide variety of systems as a network. As some of the following examples will show, an astounding number of things we see around us are essentially networks. Let us consider a few of them in order to develop an intuitive understanding before getting into the mathematical details. We also provide references below to a few relevant papers.

Let us start with three examples of connectivity networks – transportation networks (Banavar et al., 1999; Zhan and Noon, 1998), phone communication networks (Onnela et al., 2007; Deville et al., 2014), and email networks (Newman et al., 2002a). We will consider the email network to clarify our ideas. Here, people sending an email are *nodes* and emails sent by one person to another represents a *link* between them. In the terminology of networks, we can say that these two people share an *edge* between them. In a phone communication network, again people are nodes and two people making calls to each other are linked. Note that whether the links are tangible or not is not a useful concept. Some networks such as roads are more tangible than others such as emails, but that does not add anything to the network structure as a mathematical object. Building on that line of thought, one can consider a number of other such examples – railway networks (Sen et al., 2003), the Internet (Yook et al., 2002), power grid networks (Albert et al., 2004; Motter et al., 2013), and the World Wide Web (Huberman, 2001). All of these fall into the category of socio-technical networks.

We can enhance the scope of examples a bit further to understand how the actions and strategies of human beings can be interlinked. The concept of social networks is widespread and immensely useful for understanding the nature of human interactions (Newman and Park, 2003). A social network represents a group of people connected together through certain events or linkages. Much of human social behavior can be traced back to the linkages that human beings create in various forms. Examples include online communication networks such as Facebook and Twitter (Lewis et al., 2008), sexual contact networks (Liljeros et al., 2003), caste and class networks (Davis et al., 1941; Munshi and Rosenzweig, 2016), scientific collaboration networks (Newman, 2001), migration networks (Nagurney et al., 1992; Munshi, 2003), movie actor networks (Watts and Strogatz, 1998; Herr et al., 2007), terror attack networks (Galam and Mauger, 2003; Clauset et al., 2007), and citation networks (Price, 1965; Hummon and Dereian, 1989). Even animals have their social networks (Lusseau, 2003; Farine and Whitehead, 2015).

Biological and physical networks abound: for example, protein–protein interaction networks (Jeong et al., 2001), brain networks (Fornito et al., 2016; Bullmore and Sporns, 2012), nervous system networks (Pan et al., 2010), metabolic networks (Jeong et al., 2000), food web networks (Dunne et al., 2002), disease transmission networks (Riley, 2007; Valdez et al., 2018), earthquake networks (Abe and Suzuki, 2004), and river networks (Maritan et al., 1996; Dodds and Rothman, 2000).

Given this wide variety, we need a starting point to describe networks in a coherent fashion. This is important because the underlying system matters for the functionality of a network. While networks arising out of different systems may have a common structure, their properties may differ a lot depending on which system they belong to. For our purpose we will consider economic and social network although a lot of insights will carry through for other types of networks as well. For more details on technical, physical, and biological networks, one can refer to Newman (2010) and Caldarelli (2007).

Historically, the study of networks started with graph theory. One of the earliest problems was the famous Seven Bridges of Königsberg problem, which in some

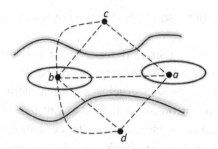

Figure 5.1 Layout of the Königsberg bridge problem. Points *a* and *b* are on two islands, points *c* and *d* are in the mainland with a river flowing in between. The arcs and lines connecting the four points show possible paths to go with one point to another via bridges.

ways gave birth to modern graph theory. Königsberg (once in Germany; now Kaliningrad, Russia) had seven bridges over a river connecting four points, two on the river's banks and two on islands in the river. The problem was to find a path to cross each of the seven bridges only once without crossing any bridge twice (Figure 5.1). Almost three centuries ago, Leonhard Euler showed that this problem does not have a solution. More importantly, his formulation provided one of the first formal treatments of graphs as mathematical objects (Euler, 1736). A major conceptual leap in his analysis was the recognition that the underlying physical locations or distances do not matter. In other words, one can move away from the physical world and consider an abstract version of the problem simply in terms of a network with four nodes and seven links. This separation provided the conceptual impetus to study the network as a separate mathematical object (Thurner et al., 2018).

In recent times, the advent of social media has made "network" a commonplace word, as reflected in the blockbuster Hollywood movie *The Social Network* (based on Mark Zuckerberg, the founder of Facebook). The study of social networks can be traced back to a pioneering analysis by Jacob Moreno, a psychiatrist. He produced one of the first systematic studies of social networks, which he conducted by mapping friendships between girls in a reform school in New York City (Moreno, 1953).

Given the astounding diversity of different types of networks that we see around us, we might ask the question: if networks are so general, is there any unifying principle or methodology that unites or finds some deep patterns in these different types of networks? Or are we limited only to descriptive devices? The answer, as we will argue in the rest of the chapter, is that we can discover both. In one form, our portrayals of networks are indeed descriptive devices that allow us to summarize information about a set of interconnected entities. This kind of abstraction is useful for mapping very different systems on to the same mathematical object. However, the theory of networks have been also very successful not only in terms of description but also in providing analytical methods for investigating causal phenomena.

Before delving into the algebraic description of networks, it is useful to take a detour through the concept of hierarchy as compared to network. Simply put, hierarchies are vertically organized structures whereas networks are more horizontally organized. We can borrow some insights from the historian Niall Ferguson to understand why, in socio-economic contexts, networks matter beyond the simple description of linkages. In his book *The Square and the Tower* (Ferguson, 2017), Ferguson argues that the current dominance of the "networked era" in public thinking is shaped by the social media revolution, which is often presented as an unprecedented phenomenon. But an earlier networked era probably followed from the invention of the printing press in Europe. These two eras were separated by an intermediate period of hundreds of years that saw the rise and collapse of hierarchies in the organization of society. Following this idea, it is tempting to conjecture that many of the socio-economic networks that we see around us have also seen a corresponding dynamic in their evolution. For a very specific example of how technological networks are related to economic and political networks, one can refer to Wenzlhuemer (2013), who discusses how the advent of the telegraph network contributed to globalization in the nineteenth century.

In this chapter we will review and discuss some basic statistics and algorithms for describing networks in a more general context, going beyond the socio-economic world. Later on, in Chapter 6, we will come back to the discussions around social and economic competitions and how linkages may or may not affect these.

5.2 Parts of a Network

In this discussion, a "network," also known as a "graph," is a set of nodes (also called vertices) connected to each other through some relationship among them (called edges or links). A network can be directed or undirected and weighted or unweighted (see Figure 5.2). In real-world networks, nodes can be, for example, people, cities, stocks, etc., and edges can be friendship, distance, similarity, etc. Notationally, we will describe a network of N nodes as $\mathcal{N} = \{V, E\}$ where V denotes a set of nodes (with cardinality N) and E denotes the corresponding set of edges. We can define two types of networks based on the directions of the edges and the weights of the edges.

It is customary to describe and/or analyze a network through its adjacency matrix. Mathematically, the adjacency matrix $\mathcal{A}$ is an $N \times N$ matrix where the i, jth element denotes the relationship between the ith node and the jth node, for all $i, j \in V$.

5.2.1 Undirected and Directed Networks with Edge Weights

As the name suggests, edges in undirected networks do not have directions. An example of such a connection is a Facebook friendship. If two people are friends on Facebook, then the *edge* exists in both ways. This is different from Twitter followers. When a person follows another person in Twitter, it is not mandatory for the person being followed to follow the follower. Thus the relationship can be asymmetric. These types of networks are called directed networks.

It is quite easy to describe directed and undirected networks in terms of adjacency matrices. But before that, it is useful to describe the weight of an edge. The elements in the adjacency matrix A_{ij} represent the weight of the edge connecting node i to node j. Weight represents a measure of the importance of an edge or a link in a network. When all edges have the same weight in a network, the network is called *unweighted*. A weighted network, on the other hand, will display variation in weights across edges. A standard way to describe an unweighted network is use binary weights such that if an edge exists between the nodes i and j (the link beginning at node j and ending at node i – we will describe the notation more elaborately below), we write $A_{ij} = 1$, else $A_{ij} = 0$.

If for all pairs of nodes $\{i, j\}$, the values in the adjacency matrix are symmetric for i to j and j to i, then the network is undirected. Otherwise, it is directed. To put it differently, if there exists at least one pair of nodes $\{i, j\}$ such that $A_{ij} \neq A_{ji}$, then the network is directed. Edges in directed networks can be differentiated as incoming and outgoing edges. Examples of real-life networks that are directed are email networks, transportation networks, and asset-holding networks.

We show some examples of weighted and unweighted networks along with directions in Figure 5.2. There are four combinations shown in four panels: unweighted-undirected, unweighted-directed, weighted-undirected, and weighted-directed. Each panel shows the plots of an example network and the corresponding adjacency matrix. Directed edges are indicated by arrows and different edge weights are indicated by varying edge widths. Each panel shows one potential node that is not connected to the network. In Table 5.1, we provide a few examples of real-life networks along with the nature of the nodes, links, and directions.

Throughout this chapter, we will consistently refer to two different types of networks. The first type is a social network, using the example of connections among company directors and corporate boards: "board interlocks." The second type is an economic network in the form of production linkages in a given country.

Bipartite Networks and Projections

In order to explain the ideas that we are going to describe in this section, we will depend on a particular type of social network. This is a specific *bipartite* network of corporate board interlock via directors (Aggarwal et al., 2020).

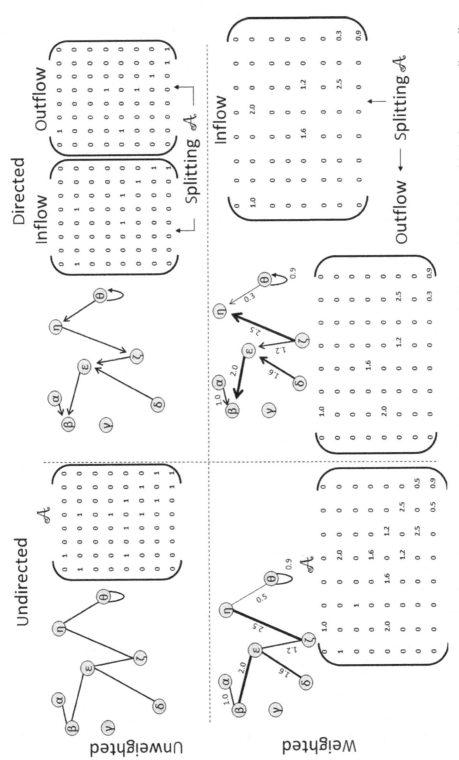

Figure 5.2 Graphical representation of networks (undirected and directed, with and without weights) along with their corresponding adjacency matrices, A. The circles represent nodes, and the lines (undirected) or arrows (directed) connecting those circles represent links or edges. The widths of the links represent the strengths or weights of the links.

Table 5.1 *Some real-life socio-economic networks with nodes, links, and direction*

Network	Nodes	Links	Direction	Reference (among many others)
Scientific collaboration	Scientists	Co-authorship	Undirected	Newman (2004a)
Social media	People	Following/ friendship	(Un-)directed	Lee et al. (2014)
Email	Email ID	Emails	Directed	Newman et al. (2002a)
Politics	People or organizations	Relations	Directed	Gil-Mendieta and Schmidt (1996)
Migration	People	Movement	Directed	Munshi (2003)
Finance	Banks and firms	Assets	Directed	Acemoğlu et al. (2015)
Markets	Buyers and sellers	Trading	Directed	Choi et al. (2017)
Innovation	Patents	Citation	Directed	Acemoğlu et al. (2016a)

Before we describe what a bipartite network is, let us see an example to form an idea about the object. In Figure 5.3 we show a schematic of a network comprising two sets of nodes – a set of company directors and a set of companies that they are on the parts of. This network is called a *bipartite* network because of its two sets of directors and boards.

This bipartite network leads to two different networks. A network can be projected only on the companies, where the nodes will be companies and a pair of companies will be connected if they share a common director; or a network can be projected only on the directors, where a pair of directors will be connected if they are present on the board of the same company. We will refer to the first one as the board interlock network and the second one as the directors' network. In what follows, we use data sampled from the Prime database (which supplies information on capital market offerings in India).

Generally, bipartite networks represent a complex set of interconnections. Specifically, they have two types of nodes and the edges always connect nodes of one type to nodes of another types, with no edge connecting the same types of nodes. Examples of such sets of nodes are movie actor networks, country trade networks, and consumer products networks, where the connections are across the sets of nodes.

The *incidence matrix* $\mathcal{I}$ describe the relationships among two types of entities. They can be nodes and edges. In this case, we can utilize them to denote two sets of nodes. If there are M nodes of type one and N nodes of type two, then $\mathcal{I}$ is an

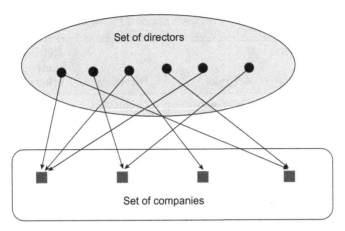

Figure 5.3 Corporate network: a schematic illustration of a bipartite network capturing director-to-company mapping.

$M \times N$ rectangular matrix. We can represent an unweighted bipartite network by an incidence matrix $\mathcal{I}$ such that if there exists a link between node i of type one and node j of type two, then $\mathcal{I}_{ij} = 1$, else 0. If each node in one group connects to all nodes in the other group, then we will call such networks *complete* bipartite networks.

5.3 Node- and Network-Level Characteristics

In this section, we will describe some fundamental network measures. These measures are generally applicable to all networks and they provide a useful set of tools and techniques to summarize global and local properties of networks.

5.3.1 Degree

Degree indicates the popularity or importance of a node in a network. Consider an undirected, unweighted network with adjacency matrix $\mathcal{A}$. The degree of node $i \in V$ is the sum of the number of links a node i has: see Figure 5.4, panel (a). Notationally, we write the degree of the ith node as

$$k_i = \sum_{j=1}^{N} \mathcal{A}_{ij} = \sum_{j=1}^{N} \mathcal{A}_{ji}. \tag{5.1}$$

The *average* degree of a network is defined as

$$\langle k \rangle = \frac{\sum_i^N k_i}{N} \tag{5.2}$$

where $\langle . \rangle$ denotes sample average. Note that the maximum possible degree of a node is $k_{max} = N - 1$, obtained when the node is connected to every other node. We

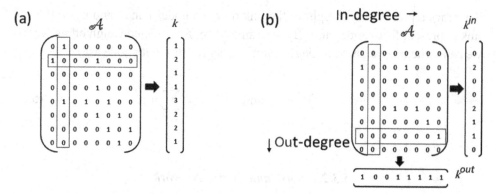

Figure 5.4 Schematic representation of in-degree, out-degree, and total degree of the unweighted networks shown in Figure 5.2 excluding the self-loop. We illustrate the connectivity of nodes through adjacency matrix $\mathcal{A}$ for undirected and directed networks: (a) total degree across nodes in the undirected network, and (b) in-degree and out-degree across nodes in the directed network. We have excluded the self-loop from our calculation to capture only the linkages across nodes.

define the density of connections as the ratio between the average and maximum degree.

As we have seen in Section 5.2.1, directed networks can potentially have both incoming and outgoing links. The number of incoming links to node i is called the *in-degree* of that node. It is important to define the adjacency matrix in a clearer way for this purpose. Technically speaking, we can use either $\mathcal{A}_{ij}$ or $\mathcal{A}_{ji}$ to denote an outgoing link from the node i to node j. In the present context, we utilize the notation $\mathcal{A}_{ij}$ to denote an outgoing link from node j to i – equivalently, that denotes an incoming link to node i from node j.

With such notation, the in-degree of the ith node is the sum of the row entries in adjacency matrix $\mathcal{A}$ shown in Figure 5.4, panel (b): that is,

$$k_i^{in} = \sum_{j=1}^{N} \mathcal{A}_{ij}. \tag{5.3}$$

The number of outgoing links (for a binary network) of node i is called the *out-degree* of that node. More generally, the out-degree is the sum of the column entries in the adjacency matrix $\mathcal{A}$: that is, for the ith node, the out-degree is

$$k_i^{out} = \sum_{j=1}^{N} \mathcal{A}_{ji}. \tag{5.4}$$

We can extend the concept of in-degree and out-degree in the case of a weighted network as well. The sum of the weights of its links defines the *weighted degree*. Therefore, the weighted degree of node i in an *undirected weighted* network is defined as

$$k_i^{w} = \sum_{j=1}^{N} w_{ji} = \sum_{j=1}^{N} w_{ij} \tag{5.5}$$

where w_{ji} represents the weight of the link between nodes i and j and $w_{ji} = 0$ if no link is present from node j to i. By the same logic, for directed weighted networks we can write the weighted in-degree and out-degree of the ith node as

$$k_i^{in,w} = \sum_{j=1}^{N} w_{ij} \quad \text{and} \quad k_i^{out,w} = \sum_{j=1}^{N} w_{ji}. \tag{5.6}$$

5.3.2 Dense and Sparse Networks

Density is a measure of how dense the connections are in a given network. There are two extreme examples. A complete network is a network for which all nodes are connected to every other node. On the other hand, a ring network is where each node is connected to exactly two neighbors such that the whole network creates a closed loop. Intuitively, a complete network is clearly more *dense* than a ring network even with the same number of nodes. How do we formally capture this idea of density? A technical way to answer this question would be to count how many links are present in the network as a fraction of the total possible links.

Intuitively, such a measure would range from 0 (no connectivity) to 1 (fully connected) after normalization. In real-world networks, the density is often much smaller than 1. The reason is that a large fraction of pairs of nodes may not have direct connections between them. If the density is very low, such networks are called *sparse* networks. The fewer edges are in a network, the sparser the network is.

The maximum number of possible links in undirected, directed, and bipartite networks are:

$$L_{max} = \begin{cases} \frac{N(N-1)}{2} & \text{for an undirected network with } N \text{ nodes,} \\ N(N-1) & \text{for a directed network with } N \text{ nodes.} \end{cases} \tag{5.7}$$

Therefore, the density of a network with N nodes and L links is

$$d = L/L_{max} = \begin{cases} \frac{2L}{N(N-1)} & \text{for an undirected network with } N \text{ nodes,} \\ \frac{L}{N(N-1)} & \text{for a directed network with } N \text{ nodes.} \end{cases} \tag{5.8}$$

In a *growing* network, the growth can be observed in terms of how the number of links increases as a function of the number of nodes. If the number of links grows proportionally to the number of nodes ($L \sim N$), then the network is sparse. If it grows faster than linearly (e.g. quadratically $L \sim N^2$) or exponentially ($L \sim e^N$) with respect to the network size, then the network is called dense.

5.3.3 Paths and Distances Across Nodes

A path from a given *source* node to a given *target* node is a sequence of edges traversed to go from the source node to the target node. In an unweighted and undirected network, the number of edges traversed in a path is called the *path length*. A given pair of nodes may have multiple paths between them. These paths may have different lengths, and may or may not share some common edges. If the source and the target nodes are the same, the path is called *cyclic*. If a path never goes through the same link more than once, we call it *simple*.

The concept of a path is the basis of the definition of *distance* among nodes in a network. A straightforward way to conceptualize distance between a pair of nodes is to consider the minimum number of edges to be traversed in a path connecting the two nodes. In other words, distance is simply captured by the *shortest path*, and its magnitude is given by the *shortest-path length*. There may be multiple shortest paths between two nodes, but they all must have the same length. The shortest paths between two nodes depend on the type of network: that is, whether the network is directed and/or weighted or not. In the case of an undirected and unweighted network, the shortest path is just the one that has the minimum number of edges. In the case of a directed network, the paths must be consistent with the direction of the links along the paths.

The *average shortest-path length* in a network is obtained by averaging the shortest-path lengths across all pairs of nodes. The *diameter* of the network is the maximum shortest-path length across all pairs of nodes, that is, the length of the longest shortest path in the network.

Mathematically, the average path length of an undirected or directed, unweighted or weighted network is defined as

$$\langle l \rangle = \begin{cases} \frac{2 \sum_{i,j}^{N} d_{ij}}{N(N-1)} & \text{for an undirected network,} \\ \frac{\sum_{i,j}^{N} d_{ij}}{N(N-1)} & \text{for a directed network} \end{cases} \tag{5.9}$$

where d_{ij} is the shortest-path length between nodes i and j, and N is the number of nodes. Using the same notation, we can write the diameter as

$$dim = \max_{i,j \in V} d_{ij}. \tag{5.10}$$

In Figure 5.5, we show two networks – one is unweighted and the other one is weighted. In both, the minimum path between nodes β and η is shown by dashed lines.

5.3.4 Connectedness of Nodes

For many processes that take place in networks, such as information diffusion (think of a rumor), the network's topology is very important. One key topological property is *connectedness*. How far information diffuses in a given social network will depend on how fragmented the network is. With a highly fragmented network, information cannot jump from one component of the network to another. However, if the network is connected in the sense that from any node one can find a path to any other node, the chances are higher that the information diffusion can actually take place.

Thus, connectedness is an important property to understand in the structure and function of a network. Note that the number of edges in a network is bounded by the number of nodes (we are ignoring the possibility of having multiple edges connecting the same pair of nodes). The upper bound corresponds to a complete network, whereas in the lower bound, a network would have no links at all. In Section 5.3.2, we saw that the higher the density, the greater are the chances that the network is *connected*. The fewer the links and the lower the density, the higher the chances that the network is disconnected, which gives rise to smaller connected *components* in the network.

A component is a connected subnetwork that is unconnected to other subnetworks (this definition can be weakened by considering sparse connections). The largest connected components in many real-world networks include a substantial portion of the network and is often called the *giant* component. In a connected network, the giant component coincides with the entire graph.

An undirected network can either be connected or unconnected. A directed network, on the other hand, can be *strongly* connected, *weakly* connected, or unconnected. In a strongly connected network, there is at least one directed path between every pair of nodes, in both directions. A weakly connected network is such that replacing all the directed edges with undirected edges will create a connected network that is undirected.

As an example, in Figure 5.6 we plot the giant component of a corporate network with a sample of 600 directors of firms registered in the National Stock Exchange a India. Two directors are connected if they belong to the same company's board. Each director can hold multiple director positions. In the figure, the giant component has 230 directors out of the total sample of 600 directors considered. We have not plotted the rest of the components as there would be too many small subnetworks to visualize. For the same reason, we do not consider the full set of directors in this figure as that would be in the order of many thousands of connections and a clean visualization would not be possible.

We show a different type of empirical network in the form of the sectoral production network in India. The data is obtained from publicly available sources on sector-to-sector flow of goods and services. Figure 5.7 shows a normalized and thresholded version of the network in 2010. This network notably has self-loops. For example, the construction sector may utilize construction services themselves.

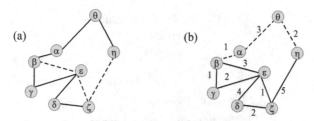

Figure 5.5 Schematic representation of path lengths in unweighted and weighted network, showing the shortest paths (dashed line) between β and η in (a) an unweighted undirected network, and (b) a weighted undirected network. In panel (a) the shortest-path length between nodes β and η in the unweighted network is $l_{\beta,\eta} = 3$: $\{\beta, \epsilon, \xi, \eta\}$; in panel (b) the shortest-path length in the weighted network is $l_{\alpha,\eta} = 3$: $\{\beta, \alpha, \theta, \eta\}$.

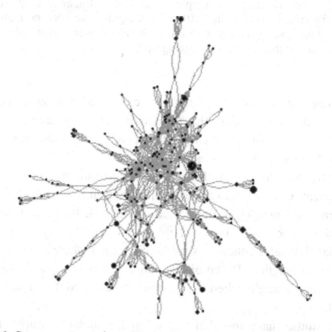

Figure 5.6 Corporate network: an illustration of the giant component of an Indian corporate network from a randomly sampled group of 600 company directors in 2016. Each node represents the director of a company registered in the National Stock Exchange of India. A link is formed when two directors are connected, that is, if they belong to the board of the same company. Single nodes have been omitted in this figure.

5.3.5 Clusters of Nodes

Transitivity is the tendency of the adjacent nodes in a network to be connected. Intuitively, it refers to a situation in a digraph in which node i is connected to node

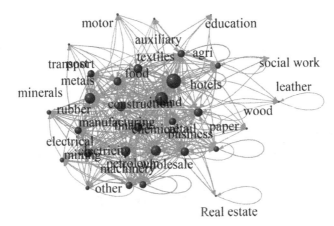

Figure 5.7 Production network: a representation of the Indian sectoral production network showing 33 sectors in 2010. Edge weights have been normalized and thresholded for ease of visualization. Node sizes have been scaled with respect to the weights of the incoming and outgoing links.

j, node j is connected to node k, and node i is connected to node k. In other words, the idea reflects the popular adage that "Friends of my friends are my friends." From the description, it is evident that transitivity depends on the existence of triads: that is, subgraphs formed by three nodes.

A quantitative way to capture this in a measure would be to consider the clustering coefficient. One can also think of clustering in the form of the likelihood of finding triplets. Consider an undirected network $\mathcal{A}$. If $\mathcal{A}_{ij} = 1$, and $\mathcal{A}_{ik} = 1$ then the nodes i, j, k form a triplet. Triplets can be of two types: triangular and non-triangular. Triangles contain six paths of length 3 whereas non-triangles contain two paths of length 2. Based on the above values of $\mathcal{A}_{ij}$ and $\mathcal{A}_{ik}$, $\mathcal{A}_{jk}$ would be equal to 1 for a triangle, whereas $\mathcal{A}_{jk}$ would be equal to 0 for a non-triangle (or open triplet).

Empirical studies have revealed that often for social networks, most of the triplets are actually triangles. To give an example, if Alagie is friends with Bikul and Chan, then it is quite likely that Bikul and Chan are also direct friends of each other. This simple observation led to the formulation of a simple measurement of transitivity of an undirected unweighted network. It can be measured simply by the fraction of all possible triplets that are triangles:

$$C = \frac{3 \times \#triangles}{\#connected\ triplets}. \tag{5.11}$$

Formally, this is called the global *clustering coefficient*. Although there were precursors to this measure, probably it is the work by Newman et al. (2002b) who made

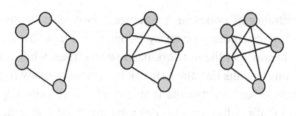

number of possible triangles increase
with more edges appearing

Figure 5.8 Schematic representation of the global clustering coefficient of three networks: $C = 0$, $15/19$, and 1, respectively.

it considerably more popular in network analysis. The value of C lies between 0 and 1 (Figure 5.8). The measure C takes the value of 1 when all triplets in the network are triangles, and C takes the value of 0 when no triangle exists.

Watts and Strogatz (1998) introduced a measure of clustering coefficient at the node level, as the ratio of existing edges with respect to the maximum possible number of edges, and at the network level, by simply averaging over all nodes. Consider a node i with degree k_i. Clearly, the maximum possible number of connections among its neighbors is

$$n_i^{max} = \frac{k_i(k_i - 1)}{2}. \tag{5.12}$$

The local clustering coefficient of node i is defined as

$$C(i) = \frac{2n_i}{k_i(k_i - 1)} \tag{5.13}$$

where n_i is the actual number of connections among the neighbors of i. Note that $C(i)$ is only defined if the degree k_i is greater than 1 (due to the $k_i - 1$ in the denominator). Intuitively, a node must have at least two neighbors for any triangle to be possible. One can construct an analogous measure for a directed network in a similar form:

$$C(i) = \frac{n_i}{k_i(k_i - 1)}. \tag{5.14}$$

Here the factor of 2 is removed since a connection A_{ij} is different from A_{ji} and, therefore, they are counted separately. We can define the clustering coefficient of a whole network as the mean of the clustering coefficients across all of its nodes:

$$\langle C \rangle = \frac{1}{N} \sum_{i=1}^{N} C_i \tag{5.15}$$

where N is the number of nodes in the network. Note that this is different from the global clustering coefficient given by Equation 5.11. A low clustering coefficient (near zero) means that the network has few triangles, while a high clustering coefficient (near one) means that the network has possibly many triangles.

As we have noted, social networks typically display a high degree of clustering. Another term that is often used to describe the same phenomenon is "*triadic closure*." Online social networks make suggestions based on these kinds of triadic closures. For example, social media platforms recommend "people you may know" based on common friends, or they may recommend accounts followed by one's friends. For the second type, we can adapt the adage of friend triplets as: "Followers of my followers are also my followers"!

5.3.6 Reciprocity of Connections

The concepts of clustering and transitivity deal with how network triplets behave, which carries information about the local topology of the nodes. However, we can look at more restricted local topology by considering *reciprocity*. The idea is to quantify how likely it is that a node with a directed link i to j also has a link in the opposite direction $-j$ to i? Reciprocity is a quantitative measure used to determine the tendency of node pairs to form mutual connections in a directed network. Needless to say, this would not make much sense in an undirected network. The measure R can be defined as the ratio of the number of links going in both directions with respect to the total number of links. For an adjacency matrix $\mathcal{A}$, we write

$$R = \frac{TrA^2}{||\mathcal{A}||} \tag{5.16}$$

where $0 \leq R \leq 1$. The case $R = 1$ corresponds to a completely bidirectional network and $R = 0$ corresponds to a fully unidirectional network. Note that reciprocal links are effectively self-loops with length 2. The numerator captures this number via squaring the adjacency matrix and taking its trace. The denominator is given by

$$||\mathcal{A}|| = \sum_{ij} \mathcal{A}_{ij}, \tag{5.17}$$

which is clearly the total number of links in a directed network. It can be confirmed that going by this formula, a symmetric directed network has $R = 1$.

Garlaschelli and Loffredo (2004) defined reciprocity as the correlation coefficient between the elements of $\mathcal{A}$, the adjacency matrix of a directed network, in the following form:

$$R_{GL} \equiv \frac{\sum_{i \neq j}(\mathcal{A}_{ij} - \overline{\mathcal{A}})(\mathcal{A}_{ji} - \overline{\mathcal{A}})}{\sum_{i \neq j}(\mathcal{A}_{ij} - \overline{\mathcal{A}})^2}. \tag{5.18}$$

The average value $\overline{\mathcal{A}} \equiv \frac{\sum_{i \neq j} A_{ij}}{N(N-1)}$ in turn is equal to the ratio of the number of edges and $N(N-1)$. The measure R_{GL} being greater than 0 indicates reciprocal networks, and R_{GL} being less than 0 indicates anti-reciprocal networks.

5.3.7 *Homophily in a Network*

So far we have been dealing with networks as being completely defined by their topology. However, often we have more information about the nodes than simply their linkages. This is especially true for social networks. For example, we see many node-wise attributes such as race, gender, age, nationality and so on. More importantly, these kinds of attributes sometimes have effects on network connectivity as well, at least at the level of pairs of nodes. A popular adage captures the underlying sense nicely: "Birds of a feather flock together."

In social networks, a known phenomenon is that nodes which are *similar* in some dimensions (i.e. share the same or similar attributes) tend to get connected to each other. This phenomenon is known as *homophily* and such mixing is often called *assortative* mixing. Examples of assortative networks abound. In citation networks, researchers cite other researchers in the same field. Websites tend to point to websites in the same language or expressing the same political views. Linguistic proximity between individuals often contributes to making friends. The assortative nature of social connections has proved to be tremendously important in designing interventions in socio-technical networks. Recommendation systems such as dating apps or software for consumable service products (e.g. video streaming and music playlists) utilize this feature by inferring matches based on correlated traits of individuals. Such correlations can emerge from geographic proximity or more abstract and less tangible measures of social proximity.

Mathematically, the idea of assortativity in networks is captured by the correlation of some pre-specified properties of the nodes (Newman, 2003). Given that assortivity is basically a measure of correlation, one can also generalize the idea to conceptualize negative assortativity. *Disassortative* networks are those in which neighboring nodes tend to be dissimilar. Newman (2003) made a very interesting observation, which seems to hold across real-world networks but there is not yet a general non-contextual explanation as to exactly why it holds true. The observation is that social networks (coauthorship networks, email networks, company directors' network, etc.) are typically assortative whereas technological networks (internet, power grid, and electronic circuits) and biological networks (protein interaction networks, food web, and neural networks) are typically disassortative. In other words, star nodes tend to connect to star nodes in the social networks, whereas star nodes tend to connect to non-star nodes in technological and biological networks. Knowing the underlying mechanism of this finding may open up new possibilities in the complex systems literature.

A simple way to characterize assortativity is to analyze node-level similarities based on degree. For example, in a network that exhibits *degree assortativity* or *degree correlation*, high-degree nodes tend to be connected to other high-degree nodes and low-degree nodes tend to have other low-degree nodes as neighbors.

Measuring Assortativity

There are two ways to measure the degree assortativity of a network, both based on measuring the correlation between degrees of neighbor nodes. The first method is based on an application of the Pearson correlation coefficient of the degree sequence. Given that it is a correlation coefficient, its value lies between -1 and $+1$ spanning from absolutely disassortative to absolutely assortative. However, this way of capturing assortativity is possible only when the trait or characteristic on the basis of which one is trying to find similarity can be quantified. Degree, for example, is clearly quantifiable. However, sometimes the trait may not be quantifiable. For example, intra-caste marriage is highly prevalent in India (as are other types of intra-cast relationships such as friendships or workplace relations). We cannot directly use the idea of degree correlations here, since caste is a qualitative idea and not quantitative *per se*. In such cases, one can use the concept of modularity.

Technically, modularity measures the degree to which nodes sharing similar traits (even if non-quantifiable) are connected. A standard approach to measure modularity is as follows. The idea is to first quantify the differences between the observed and expected fractions of edges for each pair of types, and then to sum them up. Formally, one first labels all the nodes in the network according to some pre-assigned list of labels. Given such a labeling, the modularity for an undirected network is given by

$$\mathbb{M} = \frac{1}{2m} \sum_{ij} (\mathcal{A}_{ij} - E_{ij}) \delta_{l_i, l_j}, \tag{5.19}$$

where $\mathcal{A}$ is the adjacency matrix, $E_{ij} = \frac{k_i k_j}{2m}$ where k_i and k_j are the degrees of nodes i and j, m is the number of edges, l_i is the label of the community to which node i belongs, and $\delta(l_i, l_j)$ is the Kronecker delta function which equals 1 when its arguments are the same and 0 otherwise.

The measure is such that a higher value of $\mathbb{M}$ would indicate the existence of more prominent modules. Note that the delta function serves as a filter over the edges, selecting only those edge-pairs $\{i, j\}$ whose labels denoting the traits are the same. Thus, the summation is taken only over the edges whose nodes share the same trait. The inner term is the difference between the observed fraction of all edges between i and j, which is $A_{ij}/2m$, and the expected value under a null model of a random graph with an identical degree distribution.

It is important to point out here that modularity is a common measure of community structure, an idea that we will discuss in Section 5.7. The above presentation of modularity is purely based on the concept of sharing traits at the node level. One difference between assortativity and modularity is that the former is a global macro-level measure that gives a summary statistic about the structure of the network, whereas modularity also gives a *meso*-level architecture of a network. This has a connection with the community structures discussed in Chapter 4. Such mesoscopic architecture of complex systems can be easily conceptualized through networks.

In Figure 5.9, we analyze the evolution of the directors' network. We see that as the network increases in size (both in the number of directors and in their number of board mandates), the density of connections falls monotonically and the path length increases. The clustering coefficient remains more or less similar (the fluctuation is magnified due to scaling in the y-axis) whereas assortativity increases.

5.3.8 Structural Balance

Structural balance is a concept at the level of a triad. Consider four countries: A, B, C, and D. Suppose country A has a war with country D. Country B does not endorse that and wants to make country A lose its diplomatic ties with other countries. Country C, on the other hand, has an independent stance and wants to evaluate the situation. So how would country C carry out a cost–benefit analysis? Here obviously we are ignoring all complications arising out of politics and international relations. Consider only the triad of A, B, and C. In a very simple-minded way, one can see that if country C aligns with either A or B, either way it forms a stable dyadic coalition and both countries of this dyad will have a negative relationship with the other country. On the other hand, if country C wants to maintain good relationships with both A and B while A and B are mutual enemies, then the triadic relationship would have a tendency to be unstable. This is essentially the concept of structural balance – the triad in the first case is balanced whereas the triad in the second case is unbalanced.

If we denote positive relationships with a + sign and negative relationships with a − sign, then stable configurations would have either all edges being + (i.e. + + +) or one + with two − (+ − −). On the other hand, three − (i.e. − − −) or two + and one − (i.e. + + −) would be unstable configurations.

There is a very famous result in this literature that goes by the name of the Cartwright–Harary theorem (Harary, 1953). The statement of the result is as follows. Consider a fully connected network where all edges have been assigned either

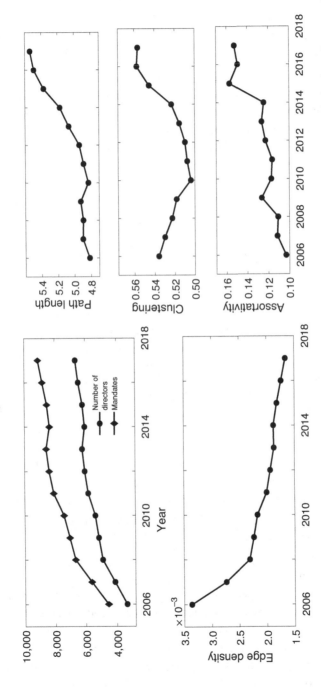

Figure 5.9 Corporate network: evolution of the company directors' network in India from 2006 to 2017. The statistics have been computed for the giant component for every year.

a positive sign or a negative sign. If the network is *structurally* balanced, then there are only two possibilities. First, all edges have positive signs. Second, the network can be partitioned into two subnetworks such that within each subnetwork, all nodes and edges have positive signs, but across these two subnetworks, all edges have negative signs.

This is a very non-trivial result. Interested readers can consult Easley and Kleinberg (2010) for a more elaborate discussion on the concept and a proof of the Harary–Cartwright theorem. There are many applications of this concept of structural balance, ranging from societal connections (Antal et al., 2006) to financial comovement networks (Kuyyamudi et al., 2019).

5.3.9 External Factors and Their Influence on Linkages

Many factors external to a given network may influence its properties. Here we describe one case based on a social connectedness network to elucidate such possibilities.

In Figure 5.10, we plot a global connectedness network across countries based on Facebook connections. The connectedness index was produced by Facebook researchers and the data is publicly available on their Data for Good website. Bailey et al. (2018) constructed this index between the ith country and the jth country, based on the following definition:

$$\text{index}_{ij} = \frac{\text{linkages}_{ij}}{\text{users}_i \times \text{users}_j}. \tag{5.20}$$

We have plotted a thresholded version of the network across 184 countries. We have also ignored self-loops. Without the threshold there would be 184^2 number of edges and the network would be incomprehensible. Equation 5.20 captures the idea of a gravity equation in trade theory which in turn is based on the Newtonian equation of gravity. In its basic form, the gravity equation states that the gravitational force between two objects is proportional to the product of the masses of the objects and inversely proportional to the square of the distance between them. Equation 5.20 applies a similar idea to find the fraction of total linkages between two regions with respect to the product of the total user base in those regions. If distance matters, then this index would vary with respect to it.

One of the most interesting findings of this analysis is that the effect of geographic distance on the social network is quite prominent. When the age of social media networks began, it was thought that the world was becoming flatter and people would become connected irrespective of distance, lessening the effects of geographic boundaries. In practice, what has happened is that the barriers due to geographic distance have reduced significantly, but the communication network as a whole has shrunk. So the relative effect of geographic distance is still there, while the nominal effect has reduced in scope and magnitude substantially.

In a very large-scale analysis based on Facebook data, Bailey et al. (2018) analyzed county-to-county social connectedness within the USA. They show that differences among counties in income, population share of whites, population share of high school graduates, religious leanings, and political leanings negatively correlate with social connectedness (Table 2, Model 5 in Bailey et al. [2018]). This finding empirically captures the effect of social homophily discussed in Section 5.3.7. This variable is also related to trade intensities, innovation intensities, and migration across counties.

This dataset represents a very rich description of the social network globally. For further work on it and on similar datasets, see the series of papers exploring various user-level demographic variables and their interactions and effects of the social network structure (Bailey et al., 2020; 2021; Kuchler et al., 2021).

This kind of analysis also opens up a new set of problems. The literature on networks in economics and the social sciences in general confronts the problem that links do not necessarily originate independently of each other. The underlying link formation process (e.g. deciding whether to connect to someone or not) is not typically observable to the modeler. This leads to the problem that links as observed may not be assumed to be exogenous. A related problem is that in social networks, people may change their behavior if they are aware of the network structure of social connectivity. In other words, links can be the result of strategic choices. There is a large literature on each of these streams of thought. Jackson (2010) and Goyal (2012) are two standard references on strategic games in networks. See De Paula (2020) and Chandrasekhar (2016) for reviews on the econometrics of networks. Breza et al. (2019) review how information flow and risk-sharing influence the formation of networks in the context of economic development.

5.3.10 Specific Subnetworks as Building Blocks

Motifs in Networks

Many real-world networks exhibit small-scale recurring patterns that appear more than would be expected by chance. These patterns in connectivity are often bigger than triangles and reciprocal connections. These structures are called *motifs*. The presence of motifs indicates a localized topological structure across a large-scale network. Intuitively, motifs are like bricks in a building, except that bricks are typically arranged in a much more coherent fashion.

Holland and Leinhardt are credited for the first systematic analysis of network motifs (Holland and Leinhardt, 1976, 1977). They introduced the idea of the

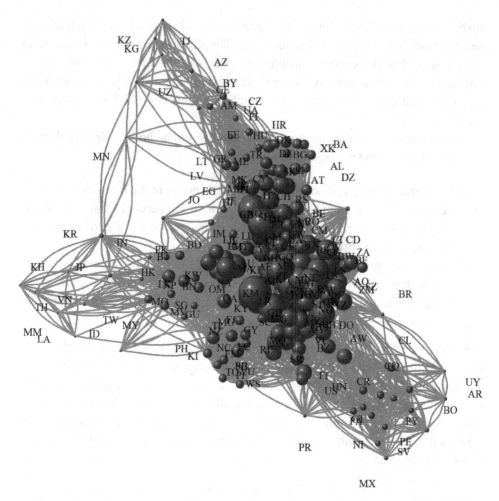

Figure 5.10 Social connectedness network across countries, based on data from Facebook. Node size is proportional to the degree of connectivity across nodes. We have applied a threshold on the edge weights for the purpose of visualization. See text for an explanation of how this representation was constructed.

statistical likelihood of subgraphs occurring in a given network (Holland and Leinhardt, 1974, 1975). The underlying idea is that if certain types of subgraphs are more prevalent than those that can be found in random networks, then they are good candidates, in a statistical sense, for being described as motifs. For a more concise description we can refer to Milo et al. (2002), who formalized this notion by defining network motifs as "patterns of interconnections occurring in complex networks at numbers that are significantly higher than those in randomized networks."

In Figure 5.11, we show some possible motifs arising out of three and four nodes in an undirected network and a directed network. In directed networks, more motifs can be generated due to the fact that the edge from node i to node j is different from the edge from node j to node i. Enumerating all motifs in a larger network (say with five nodes) will be more space-consuming as the possible number of combinations will increase more than linearly. So we will not exhibit them here. Interested readers are encouraged to try to enumerate the motifs for networks with larger number of nodes on their own.

Given the concept of motifs, the next question would be how to identify them in real networks? A standard algorithm is as follows:

(i) First, enumerate all possible subgraphs of N nodes in the network.
(ii) Perform a degree-preserving randomization of the network: that is, randomize the links while keeping the number of nodes, the number of edges, and the degree distribution unchanged.
(iii) Next, enumerate all subgraphs of N nodes of the randomized network. Subgraphs that occur significantly more frequently in the original network as compared to the randomized analog are called the "motifs" of the original network.

While the idea for motif extraction is fairly straightforward, it is quite challenging to find them. There are two major problems that crop up. First, enumerating all sub-graphs in a given network (especially if they are large) is computationally very costly. Second, assigning statistical significance to all possible sub-networks with respect to the randomized network is also difficult. However, there are many network packages in standard programming environments that are quite efficient in terms of finding motifs.

Cliques and k-Cores

Cliques are defined as subsets of nodes in an undirected network which are directly connected to each other within the subset. In other words, cliques are complete graphs which are subgraphs of a given network (Moon and Moser, 1965). The term "clique" originated in social sciences almost three-quarters of a century ago, in the study of how social connections are formed (Luce and Perry, 1949).

The concept of cliques is related to k-cores in undirected networks. These generalize the notion of degree connectivity to layers of connectivity. Formally, the k-core is the maximal set of nodes with the property that all nodes in the k-core have at least k links within the same subgraph. Mathematically, we write it as

$$\Omega_k = \max\{i | i \in V, \sum_{j \in \Omega_k} \mathcal{A}_{ij} \leq k\}. \tag{5.21}$$

If we take only the layer of Ω_k which has a degree connectivity of k, we call it a k-shell.

Undirected – three nodes   Undirected – four nodes

Directed – three nodes

Directed – four nodes

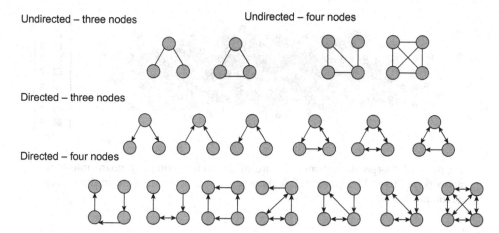

Figure 5.11 Schematic representations of key network motifs in undirected and directed networks with three and four nodes. We cannot show all possible combinations as the number of possible motifs would be too large to display here. If one allows for unconnected networks, more combinations will be possible.

By definition, no nodes with degree less than k can be in the k-core. The following steps lead to construction of k-cores:

(i) Begin with the original network and remove all the nodes with degree $k' < k$. We call this step "pruning" the network.
(ii) Remove all the remaining nodes that have degree less than k after pruning.
(iii) Repeat step (ii) until any further pruning leads to the degree of at least one residual node falling below k.
(iv) The resulting network would be a k-core or a set of k-cores of the original network.

Here it is important to point out one not-so-obvious feature of a k-core. It is *not* necessary that all nodes that have a degree more than k will be in the k-core. The reason is that due to repeated pruning, it is possible that many of the connections with nodes that do not belong to a k-core will be removed, resulting in much lesser remaining degree of a given node. If the residual degree falls below k, then that node will also be removed and cannot belong to a k-core. An easy example is to consider a star-shaped network with ten nodes, one in the center and nine in the periphery. Clearly, the degree of the central node is 9. Say we want to find the 2-core of this network. A single round of pruning of nodes with degree equal to 1 leads to pruning of all peripheral nodes in the first round. The remaining node is the central node with degree zero, and therefore that will also be pruned for $k = 1$. Thus a 2-core does not exist for this network, although the highest degree in the

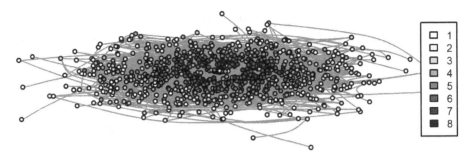

Figure 5.12 Corporate network: *k*-core of the Indian company board interlock network in 2016. The numbered scale gives the value of the core in the *k*-core decomposition.

original network was much higher than k equals 2. In Figure 5.12, we show an empirical k-core decomposition of a sample of Indian company board interlock network in 2016.

5.4 Information Content and Filtered Networks

The main reason for considering a network picture of an interconnected system is to capture the local and global topology of linkages, which presumably carries information about the nodes and/or about the network itself. However, one might ask whether it is important to consider all edges or not. The underlying point is that the simple existence of an edge may not be particularly informative. First of all, what kind of information we want depends on the goal of the analysis that we would like to carry out. Second, even if all edges carry some information, some edges may carry more information than the others. Since the number of edges increases as the square of the number of nodes in a complete network, one may want to ignore *non-informative* edges to reduce computational burden as well as to increase tractability. The question of which edge is informative depends on the definition of information and may be context-dependent. However, some useful context-agnostic toolkits have been developed in the literature that allow us to extract a subgraph from a network that preserves some properties of the original network. This exercise is often called *filtering* of the edges of a given network.

The obvious approach to filtering the edges of a weighted network is to put an exogenous threshold on the weights and remove all edges that have weights below that threshold. This kind of filtering is quite common due to its simplicity. But since its appearance in the literature, it has lost its appeal primarily due to two reasons. First, the threshold has to be context-dependent and most of the time it is not possible to come up with a threshold that works for all networks.

Second, given that the threshold values suffer from ad hoc adjustments, it is not clear *a priori* whether the filtered network suffers from other types of loss of information (or structure) or not. For example, imposing a threshold which may otherwise seem to be accurately reflecting the weight distribution (say, fixing the threshold at the median of weights), it might make the network unconnected and thus the filtered network might lose its original structure. While further ad hoc adjustments are possible, this methodology has evident theoretical and practical limitations.

With this backdrop, newer approaches have been developed that are more flexible and generalizable.

5.4.1 Extreme Filtering and Tree Structure

Trees are a special class of undirected, connected networks such that the deletion of any one link will disconnect the network into two subgraphs. The number of links in a tree with N nodes is given by $L = N - 1$. Tree networks derive their name from actual trees, a main property being that trees have *no cycles*. As a result, for any given pair of nodes, there is only one path connecting them. Trees are also *hierarchical*, that is, the structure spreads out from a root node into other sub-trees. Topologically, any node in a tree can be treated as a *root*. Each node in a tree is connected to a parent node (root) and to one or more children nodes (leaves). Two exceptions are the root node (the very first node), which has no parent, and the lowest-level leaves of the tree, which have no children nodes.

Minimum Spanning Tree

Given an undirected and connected network $\mathcal{N} = (V, E)$, a spanning tree of the network $\mathcal{N}$ is a tree subgraph that has all the nodes V and every edge in the tree belonging to E. Generally, there can be more than one spanning tree for a given network. The minimum spanning tree (often abbreviated as MST) (Cormen, 2009) is the spanning tree which has the lowest sum of weights among all the possible spanning trees.

Minimum spanning trees are useful in many different domains and have found direct applications in network design, including computer networks, transportation networks, telecommunication networks, electrical grids, and water supply networks. The theoretical importance is also immense as it has an indirect connection to the famous *traveling salesman problem*, a problem that finds the least path among a set of coordinates. While describing what is a minimum spanning tree is straightforward, how to find the tree from a given graph is less obvious. There are some well-known numerical approaches that can find the minimum spanning tree quite efficiently. Below, we describe two methodologies.

Prim–Dijkstra's Algorithm

Prim (1957) proposed a greedy approach to finding the minimum spanning tree. While the algorithm goes by the name of Prim and Dijkstra who discovered it in 1957 and 1959, respectively, Vojtěch Jarník was probably the first to discover the algorithm in 1930. This algorithm works efficiently for dense networks. Algorithm 5.1 gives a brief description.

Algorithm 5.1 Prim–Dijkstra's algorithm

---

1. Initialize an MST with a randomly chosen node on the network.
2. Collect all the edges that connect the existing tree to new nodes.
3. Find the minimum of those edges and enlarge the tree by incorporating the edge.
4. Repeat steps 2 and 3 until all nodes become part of the tree.

Kruskal's Algorithm

Kruskal (1956) proposed an alternative algorithm that also became one of the standard algorithms and that works efficiently for sparse networks, shown in Algorithm 5.2.

Algorithm 5.2 Kruskal's algorithm

---

1. Initiate a forest where all nodes are considered to be isolated trees.
2. Arrange all edges which do not belong to the MST in ascending (nondescending in case of ties) order according to their weights.
3. Choose the edge with the smallest weight. If it does not form a cycle, make it a part of the MST. Else, discard it.
4. Continue the above two steps till the tree comprises all nodes from the original network.

Note that for the first edge, there does not exist a spanning tree. The first edge itself is the origin of the spanning tree. This kind of filtering is used quite frequently with financial comovement networks. We note that this has a connection with the hierarchical classification algorithm we described in Section 4.5.2. Mantegna (1999) was one of the first papers to show the application of a minimum spanning tree on financial comovement networks and connect it to hierarchical classifications.

In Figure 5.13, we show an application of the minimum spanning tree algorithm on the sectoral production network shown in Figure 5.7. The names of the sectors are shortened for visual clarity. The original input–output network is asymmetric. We have made the network symmetric by considering the modified adjacency

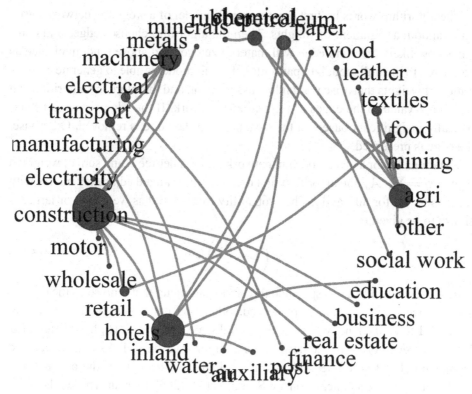

Figure 5.13 Sectoral production network: minimum spanning tree on the inverse edge weights of the sectoral production network shown in Figure 5.7.

matrix as $\mathcal{A}_{mod} = (\mathcal{A}+\mathcal{A}')/2$ ($\mathcal{A}'$ is the transpose of $\mathcal{A}$). The elements of this matrix capture the total connectivity across sectors. Before applying the minimum spanning tree algorithm, we carried out an element-wise inversion of the matrix $\mathcal{A}_{mod}$. This was done to ensure that the edges with smaller weights would be identified (subject to global optimization).

5.4.2 Statistical and Geometric Methods

Disparity Filter

Here we briefly describe the *disparity filter* proposed by Serrano et al. (2009), an important statistical filter. The baseline idea is to utilize endogenous thresholding for filtering purposes. An elementary way to apply thresholding would be to remove edges with strength less than a predetermined threshold. In contrast, here the threshold is not directly applied to the edge weight. Rather, it is determined by the statistical significance of the edges. The filtered network is shown to preserve the degree and weight distributions and the clustering coefficient of the original network.

The algorithm works by first creating a null model of a weighted network based on a random assignment of weights. Then the algorithm chooses edges and compares how likely a certain edge will materialize over and above random chance as compared to the null model. In particular, the null model is able to generate a distribution of weights that arise via random assignment, and therefore the algorithm can assign a p-value to the edges of the realized network. If the edge seems to be statistically insignificant based on the p-value, then the edge is removed. Otherwise, the edge is preserved.

Notationally, for a given node i in network $\mathcal{A}$, with degree k_i and sum of weighted degree $s_i = \sum_j \mathcal{A}_{ij}$, the algorithm first creates a normalized adjacency matrix with weights $\mathcal{A}_{ij}/s_i$ for all i and j. The probability that a link has weight w_i or larger is shown to be given by

$$p_{ij} = \left(1 - \frac{\mathcal{A}_{ij}}{s_i}\right)^{k_j - 1}. \tag{5.22}$$

If $p_{ij} < \alpha$, where α is a pre-specified threshold parameter that represents the desired significance level, the link is preserved; otherwise it is removed. Lower values of α lead to sparser networks, as fewer links are preserved. Higher values of α lead to lesser filtering of the network as a larger fraction of links will be deemed significant. The filtered network is often called the *backbone* of the original network. Interested readers can consult Serrano et al. (2009) for further details of the method.

Filtering with Planar Graphs

Tumminello et al. (2005) developed a different type of filter based on the geometry of connections. They utilized the concept of a planar graph to filter correlation networks which contain the minimum spanning tree but also contain more edges, such that triangular loops and cliques with four nodes exist. Thus the resulting structure contains a tree structure, but itself is not a tree.

Massara et al. (2017) proposed another algorithm in the form of a triangulated maximally filtered graph which approximates a weighted maximal planar graph problem. In Figure 5.14, we show an application of this algorithm to the return dataset that was used in Figure 3.4.

5.5 Influence of Nodes and Edges

In real-world networks, one can easily observe that all nodes are not equally important. Imagine a star-shaped network with one node in the center and a few nodes at the periphery. Intuitively, if the number of peripheral nodes increases, we may say

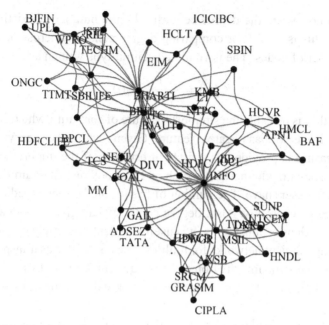

Figure 5.14 Triangulated maximally filtered graph generated from the network in Figure 3.4.

that the node at the core becomes more *important* or *influential*. However, for analytical purposes, we have to assign precise meaning to these words and *centrality* does exactly that.

Different ideas of centrality have evolved in the literature, which convey different types of information about how important or influential nodes are in a given network. Below we describe four centrality measures that are the most well-known and widely used ones.

5.5.1 Degree Centrality

Probably the most obvious way to quantify the centrality of a node is to consider how many edges does it have, the idea being that a larger number of edges should indicate a higher level of influence by the node. In Section 5.3, we labeled this quantity "degree." Following the same nomenclature, the type of centrality we are considering here is often called *degree centrality*. In the case of a directed graph, degree centrality can be defined in terms of in-degree and out-degree. Mathematically, the degree centrality $C_d(i)$ of a node i in an undirected network $\mathcal{A}$ is defined as

$$C_d(i) = k(i) = \sum_{j=1}^{N} \mathcal{A}_{ij} = \sum_{j=1}^{N} \mathcal{A}_{ji}. \tag{5.23}$$

One can also normalize the degree centrality by the number of existing nodes in the network. This is useful for comparing degree centrality across networks with different number of nodes. The normalized centrality is written as

$$C_d^*(i) = \frac{1}{N-1} C_d(i). \qquad (5.24)$$

Degree centrality is possibly the simplest measure of centrality, which is simultaneously its positive as well as negative feature. Given that it is very easy to calculate, often this measure is provided as a first-cut estimate of the importance of nodes. For example, social media networks such as Facebook or Instagram describe how many friends the users have. The number of friends on the social media platform is clearly the degree centrality. Also, degree centrality can give an idea about which nodes have very dense connections. It is useful to identify densely connected nodes when designing efficient networks for multiple purposes, such as transportation. On the other hand, the same information can be detrimental if used by a terrorist group to figure out which node to attack for maximum destabilization of the connectivity of a system.

While the ease of calculation has made degree centrality a well-known measure, for the same reason it also ignores a lot of potentially useful information. For example, by definition degree centrality accounts for connectivity only for edge lengths equal to 1. However, for many purposes (e.g. to model the spread of a disease) one may need to account for edge lengths of more than 1. In particular, neighbors of neighbors may also matter and so can their neighbors. In order to account for longer paths and more complex connections, we utilize three other measures described in the following sections.

5.5.2 Closeness Centrality

Closeness centrality (C_c) measures how *close* a node is to all other nodes in a given network. In 1950, Alexander Bavelas first defined closeness centrality (Bavelas, 1950; Sabidussi, 1966) as a reciprocal of *farness*. A more elaborate treatment of the idea was given by Freeman (1978).

The idea is as follows. For a given node, first find the shortest path (the distance) between that node and all other nodes. The closeness centrality of that node is defined as the sum of the inverse of all the distances. Sometimes it is also defined as the inverse of sum of all the distances. Mathematically, $C_c(i)$ of a node i is defined as

$$C_c(i) = \sum_{j=1}^{N} \frac{1}{d_{ij}} \qquad (5.25)$$

where d_{ij} is the distance between a pair of nodes i and j (defined in Section 5.3.3). This definition has an obvious problem that if any d_{ij} is zero, then the measure will not be well defined. To tackle such cases, an alternative definition can be considered:

$$C_c(i) = \frac{1}{\sum_{j=1}^{N} d_{ij}}. \qquad (5.26)$$

Sometimes the measure is normalized to the following:

$$C_c^*(i) = (N - 1)C_c(i) = \frac{N - 1}{\sum_j d_{ij}}. \qquad (5.27)$$

High closeness centrality is useful for percolation: for example, examining the percolation of information through a network in a short span of time.

5.5.3 Betweenness Centrality

Betweenness centrality (C_b) was originally proposed to quantify communication dynamics in a social network (Freeman, 1977). The idea is to find out which nodes act as a *bridge* between the nodes in a network. This measure captures the propensity of percolation of a piece of information from any node in the network to any other node in the network through a given node. If the given node falls frequently on the path of percolation, we assign a high value of betweenness centrality to it.

Quantitatively, the measure calculates the shortest paths among all pair of nodes and then counts how many times a given node falls on these paths. In other words, it calculates the number of times a particular node falls *in between* the shortest paths connecting pairs of nodes. The betweenness of a node i in a graph $\mathcal{N} := (V, E)$ with N nodes is computed as follows. First, for each pair of nodes (s, t), compute the shortest path(s) between them. Next, for each pair of nodes (s, t), compute the fraction of the shortest path(s) that go through the node i. Finally, sum this fraction over all pairs of nodes (s, t). Mathematically, $C_b(i)$ of a node i is defined as

$$C_b(i) = \sum_{s \neq i \neq t \in N} \frac{\phi_{st}(i)}{\phi_{st}} \qquad (5.28)$$

where ϕ_{st} is the total number of shortest paths from node s to node t and $\phi_{st}(i)$ is the number of all paths that go through i. This measure can be normalized as

$$C_b^*(i) = \frac{2}{(N - 1)(N - 2)} C_b(i)$$

$$= \frac{2}{(N - 1)(N - 2)} \sum_{s \neq i \neq t \in N} \frac{\phi_{st}(i)}{\phi_{st}}. \qquad (5.29)$$

This measure is useful for finding individuals who influence the flow of information in a network.

5.5.4 Eigenvector Centrality

Finally, we introduce the idea of eigenvector centrality, which has given rise to many other notions of centrality in the literature. This idea has become the cornerstone of network theory due to its elegant algebraic structure as well as its connections to diffusion processes in networks. We will not discuss the connection to diffusion here as it goes beyond the scope of the current chapter. It suffices here to note that this measure of centrality is a key ingredient of the famous PageRank algorithm that Google uses in its search engine. However, this measure goes far back in terms of its applications. The early ideas can be found in the pioneering work by Bonacich, who introduced an earlier version of the measure to quantify the prestige of nodes in a network (Bonacich, 1972, 1987, 1991).

The idea behind this measure of centrality is that if a given node is connected to other nodes that have higher centrality, then the given node will also have higher centrality. Note the roundabout nature of the definition. We have not described exactly what is the centrality of a given node here, but we have nonetheless described it as being related to the centralities of its neighbors. Note that a given node is a neighbor to its own neighbors. Therefore, the centrality of a given node depends on its neighbors' centralities, which in turn are also dependent on the given node's centrality. Thus, the definition creates a self-referential measure of centrality which can be solved algebraically. The name comes from the fact that the solution to this self-referential system of linear equations is given by eigenvectors of the adjacency matrix.

Mathematically, the eigenvector centrality C_e of node i in a network $\mathcal{A}$ is proportional to the sum of the centralities of node i's neighbors:

$$C_{ei} = \frac{1}{\lambda} \sum_j \mathcal{A}_{ij} C_{ej} \tag{5.30}$$

where λ is a proportionality factor. This can be rewritten as

$$\mathcal{A}C_e = \lambda C_e. \tag{5.31}$$

In the above equation, λ can be recognized as an eigenvalue of $\mathcal{A}$ and C_e is the corresponding eigenvector. One question arises here regarding which eigenvector to consider as the candidate solution. The convention is to take the dominant eigenvector to represent the centrality measure. This also derives its mathematical support from the Perron–Frobenius theorem which states that an adjacency matrix with positive real values always has a unique largest eigenvalue and that all elements of the corresponding eigenvector are also positive. This largest eigenvalue is often called the Perron–Frobenius eigenvalue. This theorem has been extended to the case of nonnegative irreducible matrices as well.

The way we have treated eigenvector centrality above is static in nature. One can also provide a dynamic interpretation of the same idea. Specifically, one can utilize the convergence of a Markov chain to capture the idea of eigenvector centrality. We will not discuss this here. Interested readers can consult Thurner et al. (2018), which provides the derivation of the centrality measure as the limit of a dynamic process.

The above four measures are the most elementary forms of centrality. In Figure 5.15, we show all four measures of centrality constructed from a subset of the Indian corporate network shown in Figure 5.6.

The literature on centrality is quite large and diverse. While it is not possible to list all measures of centrality here, we will mention two very important ones. Both of them depend on the spectral structure of the adjacency matrices (like eigenvector centrality). The first one is called PageRank. Its claim to fame came from Google's application of this measure in webpage searches and ranking. The measure builds on the idea that people searching online may hop from one page to another with some probability, and more incoming edges increase the probability of a given webpage being visited. This process can be written as a diffusion process with a damping parameter, and the resulting centrality vector is a transformation of the eigenvector centrality. The other one that has become very important in social networks is diffusion centrality (Banerjee et al., 2013). This centrality measure builds on the question of which node can act as an injection point for diffusion of information (e.g. a new product or innovation)? The more the injected information is repeated to a given node with some probability, the more likely it is for the node to be influenced by the news. Banerjee et al. (2013) show that one can construct a measure of centrality based on this idea of diffusion that is mathematically well defined under some conditions and that interpolates between degree centrality and eigenvector centrality (asymptotically). Interested readers can refer to Bloch et al. (2019) for a review of centrality measures (see also Das et al., 2018).

One important point to note here is: A natural question is whether all the centrality measures effectively convey the same information or not. The answer is – no. These measures do not generally exhibit high correlations although in some cases they may (see Valente et al., 2008; see also Li et al., 2015). The other question is how stable these measures are across samples drawn from the same population. Costenbader and Valente (2003) carried out some important works in this domain. The general consensus has been that stability of these measures is often network-dependent and, more broadly, context-dependent.

5.5.5 Robust Networks

By borrowing our intuitions about centrality measures, we can extrapolate the idea to conceptualize the "robustness" of a network. Whereas centrality captures the idea of how important one node is in a network, robustness captures the idea of how important an edge is in the network. The idea of robust networks is fundamentally related to functionalities. Consider a flow network (say, a transportation network) which allows flow of entities across the nodes through its edges. Now imagine one edge of the network breaking down. In order to ensure the same flow of entities, the load that was flowing through the edge that got broken has to be rerouted through other existing edges. But such rerouting increases loads on the residual edges and might lead to collapse of more edges, which in turn might lead to collapse of more edges due to rerouting and so on. Thus, the network might show a *cascading failure*. Robustness is defined as a property of a network in which the extent of such cascading collapse is lesser. A more general idea of robustness would be to consider the effects of removing nodes as well as removing edges. One can define robustness in a contextual way. One way to measure robustness in an abstract fashion is to consider by how much some global statistics for a network will change if a sequence of edges are removed from a given network. Typically, the size of the largest component in a network is taken to be a candidate measure for the global statistic. An associated measure is to consider the average distance between nodes in the components. To measure the extent of disruption, one starts from a given network and keeps on removing nodes, either randomly or based on hubs, once at a time. One computes the relative size of the giant component (i.e. the ratio of the number of nodes in the giant component to the number of nodes initially present in the network). Obviously, the network has to be connected at the beginning. At each time step, if the removal of a subset of nodes does not break the network into disconnected components, then the proportion of nodes in the giant component would decrease moderately. If, however, the network breaks into two or more connected components, the size of the giant component will drop substantially. As the fraction of nodes removed approaches one, a few connected components will be left in the network. Hence, the proportion of nodes in the giant component will approach zero. The literature on robust network is quite large and varied. For a mathematical model interpolating between fragile and robust networks, readers can refer to Moreira et al. (2009).

5.6 Multiple Layers of Connectivity

So far we have dealt with networks which have only one type of edge linking the nodes. But nodes might have multiple types of links. For example, individuals can

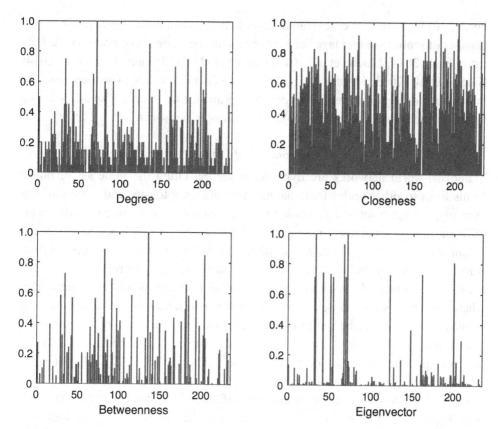

Figure 5.15 Centrality measures of the company directors in the connected component of the corporate network shown in Figure 5.6: bar plots of (top-left panel) degree, (top-right panel) closeness centrality, (bottom-left panel) betweenness centrality, and (bottom-right panel) eigenvector centrality. As can be seen in the figure, these measures may not have particularly high correlations with each other. For each of the panels, we have normalized the maximum centrality to one to scale and visualized the *y*-axis in an identical fashion.

be embedded in a collection of social networks that is a combination of linkages via friendship, family ties, coworker ties, etc. If we account for these multiple *layers* of connections with the same types of nodes but different kinds of relationships within them and across layers, then we call such networks *multilayer networks*. We show an example in Figure 5.16. This shows the trade and foreign direct investment networks across a set of European countries. The source of data for this figure is Thomson Reuters Eikon Database.

Multilayer networks consist of more than one layer of networks, where nodes in each layer are both *intraconnected* and *interconnected* (Bianconi, 2018; Demeester et al., 1999; De Domenico et al., 2013; Kivelä et al., 2014; Boccaletti et al., 2014;

Porter, 2018). A special case of multilayer networks is a *multiplex network* which consists of more than one layer of networks with the same set of nodes (F. B. Battiston et al., 2014; Mucha et al., 2010; Gomez et al., 2013; Menichetti et al., 2014; Nicosia et al., 2013; Mishra et al., 2021). In general, each layer can encapsulate information on very different types of linkages and the nodes may also differ across layers. In such cases, we call the object a *network of networks*.

Evolution of Links

Many real-world networks are dynamic in nature. Often the same set of nodes forms links to other nodes that change over time in intensity and direction. For example, the same set of Facebook friends may communicate to each other with different intensities across time. The same set of research collaborators may co-author research papers with different sub-groups of the same group of collaborators across time. In financial markets, assets may show pairwise correlations that vary across time. Intuitively, such systems can be represented as multiplex networks, where the same set of nodes keeps on evolving in terms of the nature of their connections. We describe multiplex networks that evolve following the arrow of time as *temporal* networks, where each snapshot can be interpreted as one layer of the corresponding multiplex network (Holme and Saramäki, 2012, 2019; Li et al., 2017).

Figure 5.17 shows the temporal evolution of the thresholded correlation network from Figure 3.4. Here we show two snapshots of the same set of companies over two consecutive days. The network changes as the nature of interaction changes. One way to capture the change in the network is to see the changes in some centrality measures of the nodes. In the lower two panels of the figure, we have plotted the corresponding eigenvector centralities (on the un-thresholded network; thresholding may by itself change the centrality measures drastically). The correlation measure of the 2 days' centrality vectors is 0.73.

The above exercise is generalizable. Since each of the layers of a temporal network is a static network, all measures that we have defined earlier for static networks can be applied to temporal networks with a time index. For example, the centrality measure might show time variation as dynamic links appear and disappear as opposed links changing their intensities (as in Figure 5.17).

5.7 Communities and How to Detect Them

Communities are meso-level structures of networks. Intuitively, the concept of communities in a network is related to different groups of nodes such that within-group connection density is high, but across-group connection density is low (Radicchi et al., 2004). Much of the literature around communities in networks

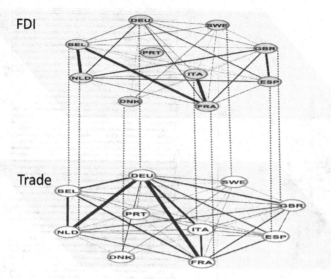

Figure 5.16 Representation of multilayer network with two layers, foreign direct investment (FDI) and trade, between 10 European countries in 2008. The nodes represent the countries and the links show FDI and trade intensities. The width of the edges represents the strength of the relationship.

has revolved around quantifying the thresholds beyond which we can call the density high (or low, for that matter, in the reverse direction). Before getting into the details, let us mention here that in what follows we do not differentiate between communities and clusters. While sometimes a distinction is made, we will follow Fortunato and Hric (2016) in describing them in exactly the same way. Therefore, in our parlance, clusters would have the same defining characteristics as communities.

Communities have two defining properties (Barabási, 2016). First, communities are connected subgraphs. Second, communities have relatively dense connections. The nodes belonging to a community have a higher probability of linking to the other nodes within that community than to nodes in other communities. Communities can be defined in the following ways:

- *Strong community:* a subgraph $\mathcal{N}^{sub}$ whose nodes are more connected within the community subgraph than with the rest of the network:

$$k_i^{in}(\mathcal{N}^{sub}) > k_i^{out}(\mathcal{N}^{sub}) \; \forall i \in \mathcal{N}^{sub}. \tag{5.32}$$

- *Weak community:* a subgraph $\mathcal{N}^{sub}$ where the sum of all degrees within $\mathcal{N}^{sub}$ is larger than the sum of all degrees toward the rest of the network:

$$\sum_{i \in \mathcal{N}^{sub}} k_i^{in}(\mathcal{N}^{sub}) > \sum_{i \in \mathcal{N}^{sub}} k_i^{out}(\mathcal{N}^{sub}). \tag{5.33}$$

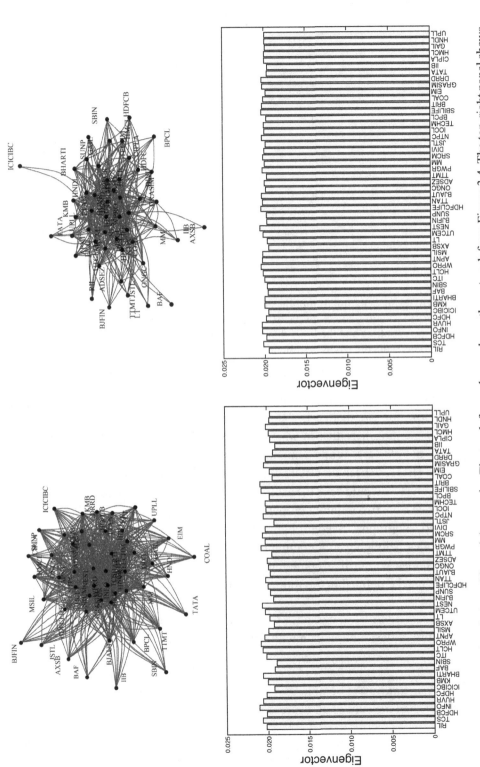

Figure 5.17 Temporal evolution of financial networks. The top-left panel reproduces the network from Figure 3.4. The top-right panel shows the network from the next day's data with the same set of companies. The network evolves with changes in the relative centrality of all the nodes. We have plotted the corresponding eigenvector centralities from the un-thresholded networks in the lower two panels.

Recalling the definition of cliques from Section 5.3.10, we note that each clique is a strong community, and each strong community in turn is also a weak community. The converse is not necessarily true. In this context, a path-breaking idea came from Mark Granovetter who distinguished between *strong* and *weak* ties in social networks. Weak ties are what connect different components of a network as opposed to the strong ties that connect within groups. Granovetter's insight was that it is the weak tie that helps in the percolation of information across social networks. This idea has become a cornerstone in network theory and goes by the name of the paper where he proposed the idea – *The Strength of weak ties* (Granovetter, 1973)!

While the idea of communities is intuitive, finding communities in large-scale networks is a challenging task. There are many methods for community detection. We will describe some of the traditional methods here. Some of the algorithms are described in the chapter on machine learning (Chapter 4), as often they go beyond networks and are more generally applicable. For reference, one should consult Fortunato and Hric (2016) for a very detailed exposition of different methods for community detection and their backgrounds.

5.7.1 Two Simple Methods

Below we describe two important techniques for finding communities. The first one is based on spectral structure and the second one is based on edge betweenness. Both are applicable to real-world homogeneous networks. Huang and Gao (2014) discuss a larger set of methods to apply to heterogeneous networks as well.

Spectral Partitioning

The first approach we will consider is called the spectral partitioning or spectral bisection method. This method is based on the spectral structure of the graph Laplacian matrix. Since the Laplacian is a fundamental construct of a network, this method gained traction in the early days of the literature when community detection *per se* was receiving relatively little attention. Later on, many better methods were developed. However, we will review this method because it is one of the most basic methods to exploit the spectral structure of a graph and it also forms the core for many other algorithms that were developed later on.

Consider a network with adjacency matrix $\mathcal{A}$ such that the network is undirected and unweighted. Let us first define the graph Laplacian as

$$\mathcal{L} = D - \mathcal{A} \tag{5.34}$$

where D is a diagonal matrix with the node-specific degrees on the diagonal. Clearly the row and the column sums of the matrix $\mathcal{L}$ would be zero. A property of the Laplacian matrix is that it will have a zero eigenvalue with eigenvector given by $1 = (1, 1, \ldots, 1)$. This observation contains the main idea of finding communities: if a network is perfectly separated in a number of communities with no cross-community linkages, then the Laplacian would be block-diagonal and each of the blocks in turn would have the same property of having zero eigenvalues along with 1 as eigenvectors of commensurate size. Therefore, for a block-diagonal Laplacian, the number of eigenvalues that would be zeroes would correspond to the number of communities. However, real networks would rarely show perfect splits and, in general, would contain some cross-community linkages. It is also to be noted that if the original network does in fact show a perfect split across communities, then separating them would be very easy and one would not require very sophisticated techniques to begin with. In case some off-diagonal links are present connecting the communities, then typically there would be one zero eigenvalue and $s - 1$ other eigenvalues corresponding to the total s number of communities, which are very small in magnitude and positive. Then one can ignore the other eigenvalues and reconstruct the Laplacian by considering these s number of eigenvalues and the corresponding eigenvectors.

While it seems quite straightforward, in practice this algorithm works well only for cases where the original network has only two communities to begin with. A larger number of communities or a relatively large fraction of cross-community linkages contaminate the results and make it quite difficult to infer the true community structure. For example, this method will completely fail in cases of correlation-based networks where the off-diagonals are non-zero and there is no evident community structure *a priori*. In Figure 5.18 we plot the Laplacian of the network shown in Figure 3.4. The off-diagonals cannot be classified easily in a binary fashion. Therefore, later developments in the literature attempted to introduce new and different tricks to bypass these problems.

The Girvan–Newman Algorithm

Girvan and Newman approached the problem of finding communities in a very different way (Girvan and Newman, 2002; Newman and Girvan, 2004). Their main idea is as follows. Fundamentally, the goal of community detection is to find dense subgraphs within a given network such that the cross-community linkages are relatively few. They exploit the idea that, as with betweenness centrality for nodes, one can create an analogous betweenness centrality for the edges. If a large fraction of the shortest paths across pairs of nodes contain a given edge, then the edge has high betweenness centrality. This immediately suggests that the edges connecting communities would have high edge betweenness centrality (relative to the ones within the communities). The algorithm is quite simple. First, one can enumerate all edges in a network and compute their edge betweenness centrality. Then the edge with

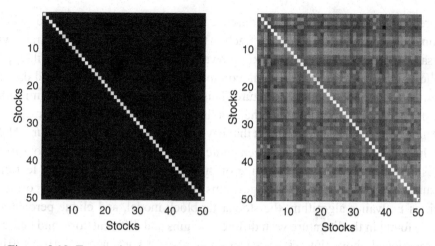

Figure 5.18 Two projections of the Laplacian of the network shown in Figure 3.4. The left-hand panel shows the color scaling with respect to the whole matrix, whereas the right-hand panel enhances the color scaling to show heterogeneity in the off-diagonal elements (darker shade indicates larger magnitude).

highest edge betweenness centrality is removed and the edge betweenness centralities of the residual edges are re-computed. If one keeps on doing this, then eventually all edges will be removed. At any intermediate stage, one can get a community structure (with varied quality). In Algorithm 5.3 we provide a pseudo-code directly borrowing the steps from Girvan and Newman (2002).

Algorithm 5.3 Girvan–Newman algorithm

---

1. Compute the betweenness centrality of all existing edges in the network.
2. Remove the edge with the highest edge betweenness. In case of ties, remove one edge randomly.
3. Re-compute the betweenness of all edges after the removal of this edge.
4. Keep on repeating the above two steps until all edges are removed.

This algorithm has a problem in that it does not give any idea as to how many communities should be there. One can stop at any stage of the edge removal process and state that the remaining network comprises the designated communities. To tackle this problem, Girvan and Newman defined an optimal community structure based on maximum modularity. Using the notation following Newman (2004b), let us define a matrix e whose generic element e_{ij} denotes the fraction of edges in the original network that connects nodes in community i to nodes in community j. Newman and Girvan defined a modularity measure Q as

$$Q = Trace(e) - ||e^2||, \tag{5.35}$$

where the Q value represents the difference between the realized fraction of within-community edges and the expected fraction that would arise in a random graph with the same degree sequence. The maximum value of modularity across possible splits of the original network into different communities provides a natural candidate for optimally chosen community structure. This principle of modularity maximization is more generally applicable than the baseline Girvan–Newman algorithm.

There are many other algorithms that have been proposed in the literature. Notably, hierarchical clustering has been quite prominent in applications to various types of networks. We describe some of them in Chapter 4 on machine learning, because the applications of these algorithms go beyond networks. Other algorithms, such as Louvain's algorithm, the stochastic block model, and clique percolation, can be found in the literature with different origins and different costs and benefits. As mentioned above, interested readers may consult Fortunato and Hric (2016), which has an in-depth review of many of these algorithms.

5.8 Network Architectures

So far we have treated networks as given and defined several metrics that quantify different features of networks. For example, for a given network, we attempted to find measures such as assortativity or degree distribution and so on. Here we flip the sequence and ask, can we create a network with some given properties? For example, can we generate a network which has a given form of degree distribution? See, for example, Figure 5.19, which shows the distribution of outflows of goods and services from the sectors we considered in Figure 5.7. The outflows are measured in millions of dollars evaluated at 2010 prices. Production networks tend to exhibit power law decay (Acemoğlu et al., 2012). In this case, such a decay is not visible since we have considered a coarse-grained sectoral decomposition with only 33 sectors. Even with a small number of sectors, we can see the non-trivial dispersion of outflow across sectors. At the level of firm-to-firm networks, the power law continues to hold (Kumar et al., 2021). Finding a generative model for such distributions is a challenging task.

Over the past few decades, scientists across various disciplines including physics, computer science, and economics have made tremendous advances in terms of generative models of networks. Here it is useful to note the key factors working behind this approach in order to understand how the networks are generated and how this relates to complex systems. Barabási (2016) summarized the focus of this development via a few major ideas. The following text is based on his descriptions of these ideas. The first point to note is that networks may occur with some given properties across many different types of systems, ranging from socio-technical to biological to physical networks.

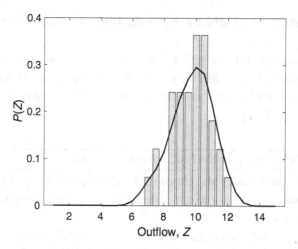

Figure 5.19 Sectoral production network: distribution of log of outflow (out-degree) with a kernel density function fitted on the empirical histogram. We have plotted the histogram, which seems to be unimodal, for the 33 sectors shown in Figure 5.7. More disaggregated datasets often exhibit a power law decay in the right tail.

Therefore, the first insight is that one may not need different network models to represent different systems. The key generative process may be identical. This lends an *interdisciplinary nature* to the models. The second feature is that while the network models can be system-agnostic, fundamentally they are driven by *empirical data*. Finally, this literature combines mathematical as well as computational approaches. Thus, taken together, these generative models can be thought of as data-driven mathematical representations of real-word networks in the form of minimum-ingredient models.

In what follows, we specify and describe three major models which are able to capture certain real-world features that occur in many commonly observed networks. We will describe random graph models à la Erdős and Rényi, small world networks à la Watts and Strogatz, and scale-free networks à la Barabási and Albert.

5.8.1 Random Graphs

This model has its origin in the work of Erdős and Rényi, who introduced a random graph model more than 60 years ago (Erdős and Rényi, 1959). A very similar model was independently introduced by Edgar Gilbert around the same time (Gilbert, 1959).

The main mechanism behind the random graph model is quite straightforward (Newman et al., 2001). This model graph can be generated in two different ways: by fixing the number of nodes and edges, which was the approach of Erdős and Rényi; or by fixing the number of nodes and treating the existence of edges as probabilistic, which was Gilbert's approach. Both give rise to the same class of networks.

For ease of description, we will follow the second approach. We start with N nodes and consider all $\binom{N}{2}$ pairs of nodes. We link each pair of nodes with some probability p. The probability parameter p lies between 0 and 1. Algorithm 5.4 shows the sequential steps.

Algorithm 5.4 Random graphs (Erdős–Rényi–Gilbert)

1. Initiate N isolated nodes and choose a threshold parameter p where $0 < p < 1$.
2. Select a node pair $N(i,j)$ and generate a random number r such that $0 < r < 1$.
3. If $r < p$, link nodes i and j.
4. Repeat steps 2 and 3 for each of the $N(N-1)/2$ node pairs.

The graph on N nodes with probability p is also known as the $G(N,p)$ model, as shown in Figure 5.20. Asymptotically, the difference between the Erdős and Rényi graph and the Gilbert graph becomes negligible. The $G(N,p)$ model has well-defined probabilistic characteristics. The first thing to note is that since the existence of edges is probabilistic, two realizations of the $G(N,p)$ model would not have the same number of edges. The expected *number of edges* is $E = \binom{N}{2} \cdot p$. The *average degree* can be calculated as $\langle k \rangle = p(N-1) \approx pN$ for large N. The expected local *clustering coefficient* is $C = p = \frac{\langle k \rangle}{N}$. The *average path length* would be $\langle l \rangle = ln\frac{N}{\langle k \rangle}$. The binomial degree distribution asymptotically converges to a Poisson distribution:

$$\binom{N-1}{k} \cdot p^k (1-p)^{N-1-k} \longrightarrow \frac{(pN)^k e^{-pN}}{k}, \qquad (5.36)$$

as $N \to \infty$ and $pN =$ constant.

5.8.2 *Small Worlds*

Erdős–Rényi networks have short paths, but triangles are rare, resulting in comparatively low values of clustering coefficients. However, real networks often differ from random graphs in this respect. In 1998, Watts and Strogatz introduced the *small-world model*, which generates networks with both features – short paths and high clustering (Watts and Strogatz, 1998; Watts, 1999). This model achieved scientific stardom because it captured the famous "six degrees of separation" phenomenon (Newman, 2000, 2003).

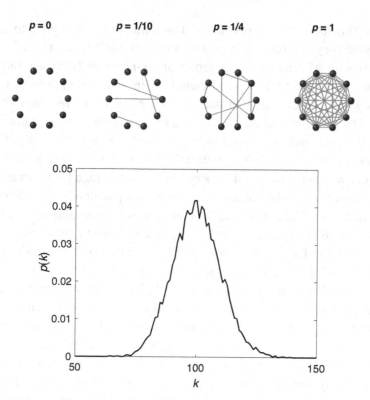

Figure 5.20 Simulation with random graphs. Top panel: evolution of random network model of 10 nodes with growing value of p (left to right). At $p = 0$, the graph is completely unconnected. At $p = 1$, the graph is fully connected and is a complete graph. For intermediate values, fraction of edges appear probabilistically. Bottom panel: simulated degree distribution of random network over $N = 10^4$ nodes and link probability $p = 0.01$, with 10^4 draws. As is evident, the average connectivity is numerically close to the theoretical value of $(N - 1)p$. The curve can be smoothed further by taking average of simulated distributions.

Let us first define the idea behind six degrees of separation. Average path length quantifies how close or far apart nodes are in a network (see Section 5.3.3). Grid-like networks typically display high average path length as compared to social networks, which often display very low path lengths. This quantitative feature was seen in the collaboration network of Paul Erdős (of the Erdős–Rényi model), an extremely prolific mathematician who collaborated with more than 500 scientists during his lifetime. Given his fame and this large number of collaborations, a new metric was developed informally known as the "Erdős number." Its value is 1 for his direct collaborators. It is 2 for direct collaborators of direct collaborators of Erdős (excluding the direct collaborators themselves), and so on. Interestingly, it was noticed that many seemingly unrelated mathematicians had very low *Erdős numbers*. This idea of small network distance was formally applied to social networks by Stanley Milgram in the 1960s (Travers and Milgram, 1967, 1977). In his experiment, Milgram chose a group of strangers in Nebraska and Kansas (two states in the USA), and asked them to help with getting a letter to a target person in Massachusetts (another state). All the participants were instructed to forward the letter to someone they knew on a first-name basis, who might be able to help get the letter one step closer to the target. At the end of the experiment, 64 letters out of 296 (roughly 21%) reached the target successfully, and their recorded path length was between 3 and 12 steps, with the average being around 6. This led to the formation of the phrase "six degrees of separation." This experiment has been replicated multiple times and the finding seems to be quite robust. In a study that came out in 2011, researchers examined around 800 million users who were actively using the Facebook platform, with around 69 billion friendships among them, and found that the average path length was 4.74 steps which translates into 3.74 "degrees of separation" (Backstrom et al., 2012; also Wilson et al., 2012).

Before getting into the description of a model for this phenomenon, we should note two points. First, the idea that social networks have a short diameter and an even shorter average path length was qualitatively noted (at least, conceptualized) by other people as well. As Backstrom et al. (2012) note, in the late 1920s, Frigyes Karinthy described the concept of *láncszemek* (Hungarian: "chain-links"), which essentially captures the idea that any two persons on the earth could be connected via no more than five contacts (Karinthy, 1929). Ithiel de Sola Pool and Manfred Kochen published a study of what they called "Contacts and influence" (de Sola Pool and Kochen, 1978), which also attempted to quantify the distance between people through chains of connections. But it would probably be fair to say that since the experiment by Milgram, and especially given the advent of social media, the observation has been established on a sound scientific basis. While the exact number of degrees of separation varies across samples and is a matter for debate, the presence of the "small world" phenomenon in social networks is now a widely accepted idea. Second, Milgram's original experiment clearly suffered from what is called *survival bias* in a statistical sense, potentially leading to the underestimation of the number of degrees of separation. The critique is that the statistic was created out of only the successful transmissions. If one accounts for the unsuccessful transmissions (the letters that never reached the target), then the resulting figure would clearly be much larger and the result would be correspondingly weaker. However, as we have just noted, the basic idea seems to have withstood the test of time successfully and has become a cornerstone finding in the social networks literature.

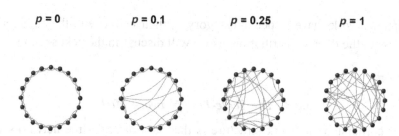

Figure 5.21 Simulation of small-world networks based on the Watts–Strogatz model: a ring network with $N = 20$ nodes where each node is connected to its immediate neighbor and to the neighbor its neighbor on each side. There are many triangles, resulting in high clustering coefficients for the nodes. However, paths from one node to another have to hop through many nodes, leading to a high magnitude of average path length. With progressively higher probability of rewiring, the network shows small-world features before eventually starting to mimic a random graph.

With the above background, let us now describe a network model for generating the "small-world" feature. The idea behind the model is to start from a grid-like network to ensure a high clustering coefficient and then to rewire some edges so that average path length is reduced (see Figure 5.21). We list the steps in Algorithm 5.5.

Algorithm 5.5 Small-world network model (Watts–Strogatz)

---

1. Construct a ring graph with N nodes, where each node is connected to its k nearest neighbors ($1 << \ln N << k << N$) with $k/2$ neighbors on both sides.
2. For each node i, rewire every edge $\{ij\}$ where j lies to the right side of node i.
3. Each edge can be replaced by any possible edge with the remaining nodes, with probability p ($0 < p < 1$) such that no self-loops or duplicate edges are formed.

The expected number of rewired links in the model is $pN\frac{\langle k \rangle}{2}$. The resulting graph interpolates between two special cases: $p = 0$ corresponds to the regular network, and $p = 1$ corresponds to a random network. For intermediate values of p, the graph can have enough shortcuts to make the distances small on average, but not so many as to disrupt the clustering behavior.

Amaral et al. (2000) provide a classification of small-world networks in large real-world interlinked systems – scale-free, broad-scale, and single-scale. Scale-free networks have degree distribution decaying as a power law. Broad-scale networks are similar except that the distribution has a sharp cut-off. Single-scale networks are those with a degree distribution that decays fast. Among these, scale-free networks have a special category of their own due to the emergence of power law in the degree distribution, as we will discuss in the next section.

5.8.3   The Scale-Free Network Model

Another cornerstone network structure is the class of scale-free networks which have power law degree distributions. This type of network exhibits hubs and spokes, where hubs are highly connected central nodes. Such structures are not seen in the other two types of networks. The hub and spoke structure is a very common occurrence in many social, biological, and physical networks, where the degree distribution shows large inequality. Some nodes may have barely any connections and some may have a much larger number.

In our description of a generative mechanism of scale-free networks, we will follow the Barabási–Albert model (Barabási and Albert, 1999) as this is probably one of the most popular models in this literature. Their model has two crucial features, *preferential attachment* and *growth*. Historically, Yule (1925) used the preferential attachment mechanism for probably the first time to explain power law distributions. Almost 50 years later, Derek de Solla Price applied the same underlying mechanism to model the growth of citation networks constructed from scientific papers (Price, 1976). Barabási and Albert rediscovered the process independently and presented a systematic treatment of the mechanism in the context of scale-free networks, as shown in Algorithm 5.6.

Algorithm 5.6 Scale-free network model (Barabási–Albert)

---

1. Initiate a network with connected nodes $n_0 \geq 2$ each node having at least one edge. For each iteration repeat steps (2) and (3).
2. At every time point, add a new node to the network forming $n \leq n_0$ new connections with the incumbent nodes with probability π which evolves at every time point. This number n is a parameter of the model and controls the average degree of the resulting network.
3. The probability π_i that a new node will be connected to incumbent node i is proportional to the degree k_i of node i:

$$\pi_i = \frac{k_i}{\sum_j k_j}. \tag{5.37}$$

Figure 5.22 shows examples of a network with power law degree distribution. Due to the preferential attachment mechanism, nodes that get more connections in

N = 10   N = 100   N = 1000

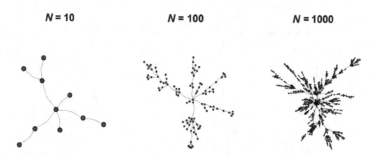

Figure 5.22 Scale-free network based on the Barabási–Albert model simulated with $N = 1,000$ nodes. The growth process is shown at three snapshots ($N = 10$, 100, and 1,000).

the beginning will continue to probabilistically dominate nodes with lower initial degree in the subsequent steps. Eventually, this leads to the emergence of large disparity in degree connectivity across nodes, which can be dubbed a "rich get richer" phenomenon. There are two distinguishing features of the resultant network. First, it exhibits hubs. Second, the process is dynamic and the degree distribution converges to a power law asymptotically. Mathematically, the asymptotic degree distribution is given by

$$P(k) \sim k^{-3}. \tag{5.38}$$

The average path length is given by $\langle l \rangle \sim \frac{\log N}{\log \log N}$ if at least two links are added at every time point, where N is the total number of nodes (Bollobas and Riordan, 2004). If a network is characterized by the existence of a relatively large number of hubs so that the average path length scales as $\langle l \rangle \sim \log \log N$, then we call it an *ultra-small world* network (see Cohen and Havlin, 2003).

5.8.4 Shocks: Exogenous Source and Endogenous Diffusion

Network models have been very useful for understanding the process of diffusion. For example, how a rumor spreads or a disease spreads through a given social network is of direct and practical importance to scientists and policy-makers alike. In the wake of the COVID pandemic, we are acutely and painfully aware of the role of transportation and migration networks in disease propagation.

Here we will discuss a particular type of shock diffusion in the context of production networks. The production network of a country can be envisaged by considering sector-to-sector connectivity or even firm-to-firm connectivity. Usually, sectoral connectivity data is easier to obtain (e.g. Figure 5.7 shows the Indian sectoral network). The methodology we are going to describe below is in sync with our discussion on time series analysis in Chapter 3 and the context is how an

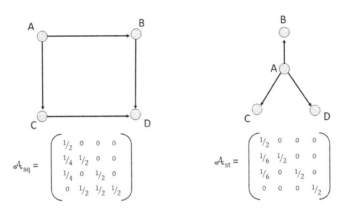

Figure 5.23 Schematic of two different networks with different implications for shock diffusion.

external perturbation at node level can spill over to neighboring nodes and their neighbors, and so on.

As a starting point we can borrow the multivariate time series framework from Chapter 3, to write down shock diffusion in discrete time through the adjacency matrices in the form

$$s_t = A s_{t-1} + \varepsilon_t. \tag{5.39}$$

Consider Figure 5.23 for two possible networks and their adjacency matrices. The dimension of the state vector s_t at a given time point t is 4×1 as we have considered four nodes in Figure 5.23. We assume that in the beginning, the system is in steady state indicating that $s_{t=-1} = [0\ 0\ 0\ 0]'$. The two cases we are considering are captured by A_{sq} and A_{st}. The first one models a square-shaped network whereas the second one models a star-shaped network. Now, we see the effect of unit shocks on the core node A in both the networks and how they diffuse over the networks via the linkages. We can simply iterate Equation 5.39 by initiating the external perturbation vector at time point $t = 0$ as $\varepsilon_0 = [1\ 0\ 0\ 0]'$, and thereafter it is a zero vector for the rest of the time, in both cases. Specifically, $\varepsilon_t = [0\ 0\ 0\ 0]'$ for $t \neq 0$.

We plot the resulting impulses across the nodes in Figure 5.24. Specifically, we plot s_t^i over time points t from 0 to 10, for all four nodes – $i \in \{A, B, C, D\}$. Clearly, the network topology plays a role in determining the extent and mode of shock diffusion. In the left panel (where the adjacency matrix is given by A_{sq}), node D responds a little later than B and C as it does not directly receive the shock, and when it does, its response is larger than the rest as it receives two shocks (from B and C). In comparison, the responses are exactly identical across B, C, and D in case of the star graph (with adjacency matrix given by A_{st}). This is expected given the symmetry of these three nodes around the core node A.

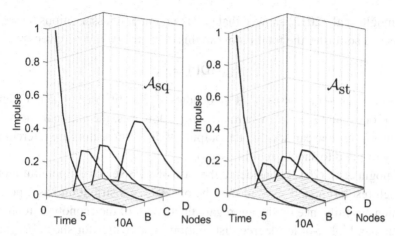

Figure 5.24 Impulse response function arising out of a shock to the core node A in Figure 5.23.

5.8.5 A Framework with Optimization

The above kind of framework is somewhat simplistic as a model for a production network. We have not considered the origin of the adjacency matrices and we also did not talk about whether the production units (firms or plants) can optimize their choices or not. Acemoğlu et al. (2012) proposed a framework which accounts for these deficiencies and has become a cornerstone model in this literature. We will now briefly review their model.

Imagine an economy defined by a production network $\mathcal{N}$. The network has N sectors (e.g. we considered 33 sectors in Figure 5.7). Assume that there is a continuum of consumers with utility function over consumption of these N goods, given by

$$\mathcal{U} = \Gamma \prod_{j=1}^{N} (c_i)^{1/N} \tag{5.40}$$

where c_i denotes the consumption of the ith sector's good. Each sector has a production function given by

$$o_j = z_j^\alpha l_j^\alpha \prod_{i=1}^{N} o_{ji}^{(1-\alpha)w_{ji}} \tag{5.41}$$

where the jth sector produces output o_j by combining productivity shock z_j, labor employed l_j, and a combination of goods purchased from all sectors $i \in \{1, 2, \ldots, N\}$. The out-degree sequence in this economy is given by $\{d_j\}$ where each element is given by $d_j = \sum_{i=1}^{N} w_{ij}$. Let us define a linkage matrix W by collecting all w_{ij}. Both consumers and producers in different sectors optimize, subject to their respective resource constraints.

Acemoğlu et al. (2012) showed that under a number of assumptions (competitive markets and so forth), the total output in this economy would be given by

$$\log(\text{GDP}) = v'\varepsilon \qquad (5.42)$$

where $v = \frac{\alpha}{N}[I - (1 - \alpha)W']^{-1}\mathbf{1}$ (I denotes an identity matrix and $\mathbf{1}$ denotes a vector of ones) and $\varepsilon = \{\varepsilon_i\}$ where $\varepsilon_j = log(z_j)$, that is, the external shocks at the node level. Thus one can see that the outputs of different sectors are interconnected and may respond to shocks to other sectors.

Acemoğlu et al. (2012) considered the case of a US sectoral input–output matrix with high resolution and showed that the out-degree distribution has a power law tail which has influence on the aggregate fluctuation of the economy's total output. In particular, higher-order degree distributions also matter for shock propagation. Interested readers can consult the paper for more economic insights (see also Acemoğlu et al. [2016b] for a review). While this is a static model, the idea of shock diffusion across sectors is embedded in this framework with optimization.

5.9 Taking Stock and Further Reading

There are several excellent textbooks and its monographs on network theory and applications. Interested readers can consult Menczer et al. (2020), Wasserman and Faust (1994), and Watts (1999). Newman (2010) presents a comprehensive description of the standard toolkit of network theory with an in-depth overview. Easley and Kleinberg (2010) provide a nice overview from an economic point of view. Sen and Chakrabarti (2013) present a complementary discussion from the point of view of sociophysics. Probably the reference most relevant to our discussion here would be Thurner et al. (2018) who describe networks in the context of complex systems.

Over the past couple of decades, the literature on networks has seen explosive growth with many new theories, application, and extensions. The study of networks has provided a useful conceptual toolkit to characterize crisis-prone systems such as financial systems and has provided new ways to evaluate systemic risks (e.g. see Battiston et al., 2012; Battiston et al., 2016; Kuyyamudi et al., 2019). Direct descendants of this literature focus on improving the resilience and robustness of networks in the face of external perturbations. A nascent literature in economics studies how information percolates in a given social network and is being actively used to formulate policy interventions to carry out targeted informational campaigns (Banerjee et al., 2013). Network theory has provided a different approach to understanding how scientific progress occurs (Acemoğlu et al., 2016a) and how that leads to paradigm shifts, eventually influencing the growth and productivity of the global economy.

Some of the latest applications of network theory to the complex systems literature can be found in Sinha et al. (2010) and Chakrabarti et al. (2019). A recent stream of work in the network literature has focused on networks of networks. D'Agostino and Scala (2014) provocatively described this as the last frontier of complexity. The network view is not only useful, but has far-reaching implications as one of the pillars in the complex systems literature (Barabási, 2016).

Part IV
Emergence: From Micro to Macro

6

Interaction and Emergence
Agent-Based Models

One of the key learnings from the complex systems literature is that simple microscopic interaction rules lead to the emergence of directly observable non-trivial macroscopic behavior that is not manifest in individual constituent units. This statement has two implicit but essential components. First, macroscopic behavior is easier to quantify and match with the observed behavior of real-world systems. Second, one can, at least in principle, construct microscopic interaction rules, that is, the rules that govern the behavior of the smallest components of the complex system, which potentially give rise to such macroscopic behavior. The prior chapters in this book provide the necessary toolkit to characterize the explicit and often implicit patterns in the data arising out of complex systems at the macroscopic level. In this chapter, our main motivation is to study *microscopic interaction rules* that lead to observed *macroscopic behavior*. To begin with, we note that in many cases such macroscopic behavior emerges in the form of a very specific feature – power law (Gabaix, 2016; Newman, 2005). Therefore, studying the emergence of power laws will be a key concept in much of our discussion in this chapter.

In what follows, we will go through a sequence of models arranged in terms of ascending complexity of the nature of interactions and heterogeneity in the resulting system-level behavior. The goal of this chapter is to give the reader a bird's-eye view of the literature of agent-based models in the context of complex systems. In each section, we will present a system which exhibits emergent behavior in one form or another. Then we will present a model to capture that system's static and dynamic properties, and discuss the scope of the modeling approach and some alternative routes that have been explored. While selection of models and their formal treatments are not exhaustive, they provide a reasonably large coverage of complex socio-economic systems. For a more in-depth as well as wider view of agent-based modeling approaches, readers can consult Namatame and Chen (2016), Chakrabarti et al. (2019), and the references therein.

6.1 Social Segregation: Interactions on Grids

Social systems exhibit many intriguing dynamical properties. To mention a few – fashions come and go, opinions swing rapidly from one extreme to the other, memes spread like wildfire, and societies remain segregated for centuries even though people may generally agree on the idea that assimilation may in fact provide tremendous economic benefit. While each of these topics has attracted attention, we focus on the idea of segregation as a starting point, for two reasons. First, the persistence of segregation along the line of visible human characteristics has perplexed researchers for a long time. Second, it also provides a way into the modeling of complex systems by the virtue of the simplicity of its benchmark model.

Here we focus on the Schelling model (Schelling, 1971; Namatame and Chen, 2016) which is generally considered to be the simplest model of spatial interaction. The main idea is that if people have preferences for visible attributes and they have some mobility, then segregation along the line of the attribute occurs very rapidly. It is not so surprising that if people can move around and they strongly dislike staying in close proximity of people belonging to other groups, then the society will be segregated. What is surprising is that even a small amount of in-group preference leads to wide divergence and segregation.

Formally, let us imagine a square grid of size $N \times N$ with N^2 cells. All the cells are considered empty at the beginning. There are two types of agents. Let us call them *red* and *blue*. We initiate M number of agents, where $M < N$. Initially, the agents are randomly distributed on the cells. Each cell can be occupied by only one agent and each agent can occupy only one cell at any given point of time. All agents like to be surrounded by other agents of the same color: that is, red agents would like to have the fraction of neighbors who are red to be greater than some number θ and blue agents would also similarly like to be surrounded by at least θ fraction of neighbors being blue. Clearly, θ is a fraction between zero and one. In principle, we can make this *tolerance* parameter θ to be agent-dependent (one can alternatively define the tolerance parameter as $1 - \theta$ which captures the tolerance for out-of-group agents; however, the mechanism and the core results remain identical). To keep it simple, let us assume that all the agents have the same tolerance level θ. Note that in a square grid, the number of neighbors of an agent is eight (right, left, front, back, and four corners). At every time point, each agent counts the number of neighbors of each color and checks whether the ratio of agents belonging to the same group as the focal agent with respect to the total number of neighbors is more than θ or not. If it is more than θ, then the agent stays put. If not, the agent moves around and finds an empty cell at random. This process continues in discrete time.

The result of this model is very surprising. Note that for large values of θ, the agents have more incentive to migrate (high in-group preference). Thus, it is intuitive that they would move out of mixed neighborhoods, which eventually would lead to segregation. The interesting result is that even for relatively small θ (which indicates a high level of tolerance for diversity), segregation emerges very quickly. We provide the pseudo-code for this in Algorithm 6.1.

Algorithm 6.1 Segregation model

Input: Set of two types of agents $\mathcal{A}_r$ and $\mathcal{A}_b$, endowed with a tolerance parameter θ; a square grid of size $N \times N$.

Output: Allocation of agents across the grid and emergence of segregation.

1. Allocate two sets of agents $\mathcal{A}_r$ and $\mathcal{A}_b$ randomly across the cells of the $N \times N$ grid.
2. All agents are endowed with a tolerance parameter $\theta \in (0, 1)$.
3. Time is discrete. At every time point, each agent calculates the fraction of agents with the same feature (red/blue) in its immediate neighborhood.
4. If the fraction falls below θ, the agent is *unsatisfied* and randomly selects and moves to an empty cell on the grid.
5. Repeat step 4 till all agents are *satisfied* with their neighborhoods.

6.2 Ripples on Sand-Piles: Self-Organized Criticality

For a long time, it has been conjectured that many natural and socio-economic systems display self-organization. A more peculiar behavior in the form of self-organized criticality (SOC) is seen in natural systems. This theory provides a good model for various complex systems: for example, earthquakes, mountain ranges, river networks, coastlines, human violence and wars, stock market fluctuations, and computer networks (Roberts and Turcotte, 1998; Picoli et al., 2014; Stauffer and Sornette, 1999; Malamud et al., 1998; Babcock and Westervelt, 1990; Elmer, 1997; Yuan et al., 2000; Garrido et al., 1996; Sapozhnikov and Foufoula-Georgiou, 1997; Diodati et al., 1991; Bak and Chen, 1991). The intuition behind it relies on capturing *bursty* behavior, where a system slowly builds toward instability before exhibiting fluctuations of large magnitude and then reverses back to a state of relative tranquility. Technically speaking, the SOC models consist of two essential features – the presence of '$1/f$' noise in time and scale-invariant self-similarity.

Here we describe the famous Bak–Tang–Wiesenfeld (BTW) model, often called the sandpile model, which captures "$1/f$" noise (Bak et al., 1987, 1988; Bak and

Tang, 1989). We will describe the baseline model below for simulation purposes; interested readers should consult the cited references for a more in-depth understanding of the model (Bak et al., 1987, 1988; Bak and Tang, 1989; Dhar, 1990, 2006; Namatame and Chen, 2016). In particular, Dhar (1990, 2006) provides a nice exposition of the theoretical structure of the problem and the emergence of universal behavior. We note that the baseline model we describe here is one specific example of the general class of models. Several variations are possible based on what is described below, which we omit as they fall outside the scope of this chapter.

The main idea is as follows. Consider a cellular automata model of sandpile growth. Let us start with a two-dimensional square lattice of finite size (often described as a mesh or grid) with some randomly distributed *sand grains* on it. Each cell in the grid has a carrying capacity h_c. Let us denote the height of a sand column for the ith cell on the grid by h_i. Time (t) is discrete. At every time point t, a single sand grain is added to a randomly chosen cell, increasing the height of the sand column in the cell ($h_{j,t} \rightarrow h_{j,t} + 1$ for some jth cell). If all sand columns in every cell have heights less than the carrying capacity, then the system remains dynamically stable. However, once a sand column goes beyond its carrying capacity, it collapses and some grains spill over into the neighboring cells. But due to such spillover, it is possible that some of the neighboring cells then attain column heights higher than the carrying capacity. In that case, they also collapse and the spillover continues until every cell settles down with $h_i \leq h_c$ for all i. Note that during such spillovers, it is possible that sand particles will fall over the edge. The number of sand particles cumulatively toppled at a given time point is called the "avalanche" size.

The interesting behavior of the model is that for a long time, the sandpile may not show a lot of adjustments in height. But once the average height reaches close to the system-level critical point, the addition of even a single sand particle might create a large avalanche. In the presence of such bursts, the cause–effect relationship is not proportional in this model. The size distribution of the adjustments exhibit a high degree of heterogeneity characterized by extreme fluctuations with a power law decay, similar to the ubiquitous "$1/f$" noise. Statistics of the size and duration of these avalanches have well-known scaling properties (Dhar, 1999; Ivashkevich and Priezzhev, 1998). The key insight coming out the model is that a system without any fine-tuning from the outside can endogenously generate large-scale fluctuations displaying scaling behavior. We show the outflux of sand grains in a simulation of the model in Figure 6.1 (Bak and Paczuski, 1995).

Algorithm 6.2 Self-organized criticality

Input: A two-dimensional grid $\mathcal{G}$ of size $N \times N$; a sequence of *sand* particles; denote the height of the ith cell at time t by $h_{i,t}$.

Output: Sequence of avalanches.

---

1. Choose any cell j with uniformly distributed probability and add a sand particle to it resulting in $h_{j,t+1} = h_{j,t} + 1$.
2. If $h_{i,t+1} < 4$ for all i, then the system remains unchanged.
3. If $h_{i,t+1} \geq 4$ for some i (e.g. in cell j in the first step), then $h_{i,t+1} \rightarrow h_{i,t+1} - 4$ and distribute these four sandgrains across the four neighboring cells equally. If there are no cells to distribute the spilled-over sand particle (if it is on the boundary), then the grain goes out of the grid resulting in an avalanche.
4. A single toppling can initiate a chain of toppling (due to distribution of the toppled particles among the neighboring cells). Update the sand columns till all possible topples have been accounted for and calculate the total size of avalanche and denote it by $a(t + 1)$.
5. Go to the first point above.

A follow-up model was proposed by Bak and Sneppen (1993) in the context of the self-organization of ecological systems into critical states. This model was

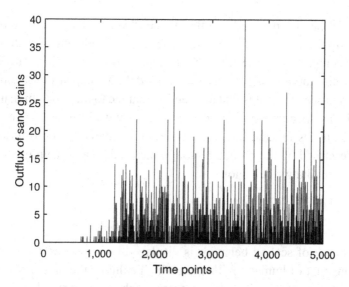

Figure 6.1 The outflux of grains falling off the grid with respect to time. The data shown in this figure was generated by simulating the BTW model of sandpile on a square lattice of size 25×25 for 5,000 time points where each time point admits the addition of one sand grain.

motivated by the finding that extinction events in the biological history of the earth are also episodic (Raup, 1986). In particular, Newman (1996, 1997) showed from fossil records that species extinction also follows the power law behavior.

This model captures the idea of "survival of the fittest." The main idea is that the model starts with N number of species placed on a ring. To fix ideas, let's say the N nodes are equidistant on a ring of some length. The exact length is unimportant here; what matters is the sequence of neighbors (left and right) for each node. The way this model works is very simple. Each species starts with some random level of fitness (fitness has uniform distribution between 0 and 1). Let's denote this by $f_i(t)$ at time t for the ith species. The initial configuration is given by a vector $\{f_i(0)\}_{i=1,...,N}$. At every time point, choose the species with the lowest fitness and replace that species along with two of its neighbors (on the left and right of the focal species) by three new species with three randomly chosen fitness levels. Then carry out the same type of pruning of the species with the lowest fitness and their geometric neighbors over time.

In each step, the system (ecology of species) updates via the *mutation* in the spirit of Darwinian evolution. However, the model introduces a nuance via the geometric linkage of the neighbors such that luck matters. Having an unfit neighbor may cause a fit species to disappear. Due to the systematic pruning, the fitness of the surviving species increases over time and reaches a critical level. Interestingly, this system is also independent of initial conditions and exhibits self-organization via punctuated equilibrium.

This model has been applied to the widely diverse mechanisms ranging from bacterial evolution (Bose and Chaudhuri, 2001; Donangelo and Fort, 2002) to competition of companies in a stock market (Cuniberti et al., 2001; Yamano, 2001). However, while the abstract nature of the model has helped in its application to many different scenarios, it is that abstraction that led to the problem that the model does not fit any specific empirical data particularly well. In fact, the experimental data on avalanches in sand-piles also do not quantitatively conform to the exact behavior described by the model, although the model has shown great success in reproducing qualitative behavior.

6.3 Size of Cities: Scaling Behavior

A different type of scaling behavior is displayed by city sizes. Cities represent a modern construct of human civilization which exhibits the properties of complex systems. In most cases, cities grow spontaneously without top-down explicit planning by a social planner or a government. Urban agglomeration leads to bursts in social and economic activities. For example, the GDP of greater New Delhi (the capital city of India) dwarfs the GDP of entire states in India. Larger cities such as New York, Shanghai, and London produce GDPs that are almost equivalent to several countries' total production. Even within countries, such activities are extremely concentrated in only a handful of regions. This pattern holds not only in India, the USA, and China, but also across most of the countries in the world. This is often called the 80–20 rule where 80% of outputs can be attributed to 20% of inputs for a given event. A more quantitative version of the rule, the Pareto principle, describes a power law behavior. Initially, the economist Vilfredo Pareto showed the power law in the context of wealth distribution (a feature that we will refer to in Section 6.4). Such behavior was also recorded in human behavior. Notably, the linguist George Zipf showed that an inverse relation exists between the frequency of usage of a given word and where it ranks among all words in terms of usage. This observation can be quantified by a power law with coefficient 2, which came to be known as Zipf's law:

$$P(S) \sim S^{-(1+\alpha)} \tag{6.1}$$

where α takes the value of 1 for Zipf's law. In this chapter, we introduce Zipf's law in the context of city size distribution, building further on the nature of the complexity of agent-level interactions described in earlier chapters. Zipf's law is ubiquitous in economic contexts even beyond city size. Axtell (2001) documented that US firm sizes follow Zipf's law.

The empirical literature has widely used the rank-order regression approach to examine Zipf's law for cities in the form of

$$\log(R) = A - \alpha \cdot \log(S), \tag{6.2}$$

where R represents the rank of the city, A represents a scalar, S represents the size of the city, and α is the coefficient of interest. This coefficient can be estimated using the ordinary least squares (OLS) estimator (see Section 2.3.5). The estimated value of α being statistically equal to one indicates the existence of Zipf's law. Following our earlier discussion on linear regression in Chapter 2, one can utilize a standard t-test with the null hypothesis that $\hat{\alpha}$ (the estimated coefficient) is not different from one.

Although intuitive, the above approach suffers from statistical inaccuracies. Gabaix and Ioannides (2004) argued that the estimated value of α from Equation 6.2 is likely to be downward biased due to the effects of outliers, especially in small samples. Additionally, they argued that the standard errors for α would also be underestimated. Consequently, regular hypothesis testing would not yield correct inferences. Gabaix and Ibragimov (2011) proposed a modification of Equation 6.2 by introducing the log of rank – 1/2 as a dependent variable,

$$\log\left(R - \frac{1}{2}\right) = A - \alpha \cdot \log(S), \tag{6.3}$$

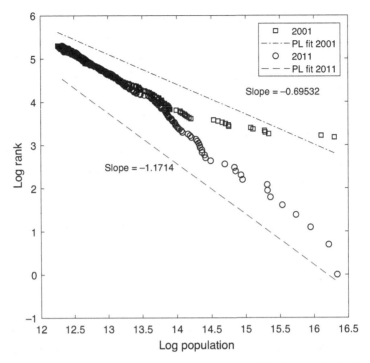

Figure 6.2 Empirical rank-plot of the top 200 Indian cities by population in 2001 and 2011 along with the power law fits (on a log–log plot using Equation 6.3).

which accounts for the bias in the estimator. Figure 6.2 shows the empirical rank-plot of the top 200 Indian cities by population in 2001 and 2011 along with the power law fits. The data was obtained from Indian Census reports. Clearly, the distributions have changed over time and the distribution shown for 2011 is closer to Zipf's law.

6.3.1 Gibrat-Type Dynamics and Zipf's Law

Here we describe a generative model of Zipf's law based on the celebrated dynamical growth model proposed by Gibrat (1931). The central mechanism for explaining the scaling behavior is the *proportional random growth* model. We will describe the main idea below, by and large following the derivation described by Gabaix (1999).

Let us consider an economy with a continuum of cities that evolves in discrete time (i.e. $t = 0$, 1, 2, and so on). Gabaix used population as a proxy for city size. Here we denote the size variable by a generic variable S. Let S_t^i be the size of city i and $\overline{S}_t$ the average size at a generic time point t. We define $X_t^i = S_t^i / \overline{S}_t$

as the normalized size. This normalization is necessary to ensure a steady-state distribution of cities.

Now, suppose each city i has a normalized population of X_t^i at time t. Let us define a proportional growth mechanism by which the cities expand as follows:

$$X_{t+1}^i = \xi_{t+1}^i X_t^i. \tag{6.4}$$

Assume growth rates to be IID, with density $f(\xi)$ at least in the upper tail. This mechanism is reminiscent of Gibrat's law for firm growth. Note that the proposed mechanism does not explain exactly why the growth rate has to be randomly distributed and independent across time and cities. But it is a useful stochastic process from the point of view of generating a power law, as we will describe below.

Let $F_t(x) = P(X_t^i > x)$ which is the CCDF of city size. The F_t will evolve as

$$F_{t+1}(x) = P(X_{t+1}^i > x) = P\left(\xi_{t+1}^i X_t^i > x\right) = P\left(X_t^i > \frac{x}{\xi_{t+1}^i}\right)$$

$$= \int_0^\infty F_t\left(\frac{x}{\xi}\right) f(\xi) d\xi. \tag{6.5}$$

Now if it does have a steady-state distribution F, it will satisfy

$$F(X) = \int_0^\infty F\left(\frac{x}{\xi}\right) f(\xi) d\xi. \tag{6.6}$$

Using the functional form $F(X) = k/X^\zeta$, where k is some constant gives $1 = \int_0^\infty \xi^\zeta f(\xi) d\xi$: that is,

$$E[\xi^\zeta] = 1. \tag{6.7}$$

Therefore, if the steady-state distribution is a power law in the upper tail, then the exponent ζ is the positive root of Equation 6.6. This idea actually goes back to Champernowne (1953). Finally, note that the average size $\bar{X}$ would be constant in the steady state. Taking the expectation of Equation 6.4, we get $\bar{X} = E[X_{t+1}] = E[\xi] E[X_t] = E[\xi]\bar{X}$. Therefore, in equilibrium, we will have $E[X_{t+1}] = E[X_t]$. Since the expected growth rate is 0, one can see that $E[\gamma] = 1$, which in turn implies $\zeta = 1$ is a solution: that is, the Zipf's law satisfies Equation 6.7. The approach proposed by Gabaix (1999) for deriving this condition is to consider an essentially non-interacting system described by Equation 6.4 along with perturbations (or a reflective barrier). For the same purpose, we will follow a different approach below.

6.3.2 Interactions and City size Dynamics

Here we describe a simple interaction-based model proposed by Chakrabarti (2017) which is a generalization of the framework described in Section 6.4. The main idea is to begin with N number of cities with population given by $\{p_i(t)\}$ for $i = 1, 2, \ldots, N$. Time t is discrete ($t = 1, 2, \ldots$). At every time point, some fraction of people leave their own cities and decide to migrate. The way the model works is by imagining a central location where all possible migrants congregate and then with some probabilities, they go to all the different cities. This model is conservative in the sense that no one leaves the system and no one gives birth. One can introduce such modifications without much of a problem. But we will ignore them here to keep the model tractable and solvable. Specifically the model is defined as follows:

$$p_1(t+1) = \lambda_1 p_1(t) + \epsilon_{1t} \sum_{k=1}^{N} (1 - \lambda_k) p_k(t)$$

$$p_2(t+1) = \lambda_2 p_2(t) + \epsilon_{2t} \sum_{k=1}^{N} (1 - \lambda_k) p_k(t)$$

$$\vdots$$

$$p_N(t+1) = \lambda_N p_N(t) + \epsilon_{Nt} \sum_{k=1}^{N} (1 - \lambda_k) p_k(t) \qquad (6.8)$$

where at every time point t, the random allocation vector $\{\epsilon_{jt}\}$ satisfies an equality that

$$\epsilon_{1t} + \epsilon_{2t} + \cdots + \epsilon_{Nt} = 1 \qquad (6.9)$$

for all t. This ensures that the total mass of population across cities follows the law of conservation:

$$\sum_{i}^{N} p_i(t+1) = \sum_{i}^{N} p_i(t). \qquad (6.10)$$

We can assume that the λ parameter is uniformly distributed among agents (but fixed over time for a given agent). The remaining ingredient of the model is the construction of ϵ. Clearly, ϵ has to be positive and has to obey Equation 6.9. However, we also want ϵ to be distributed identically with mean equal to $1/N$. Chakrabarti (2012) proposed an algorithm to find the distribution of ϵ satisfying all three conditions. First, generate a vector of independent random variables $\{\xi_j\}$ drawn from uniform distribution between 0 and 1. Second, take log of those variables and multiply by -1 to generate a sequence of variables as $\{-\log \xi_j\}$. Finally, normalize each element of this vector by the sum of all elements in the vector and call it ϵ: that is,

$$\epsilon_i = \frac{-\log \xi_i}{\sum_{j=1}^{N} -\log \xi_j}. \tag{6.11}$$

For a given time period t, the above calculation generates the vector $\{\epsilon_{jt}\}$. For $t+1$, one can generate a new draw in exactly the same way and so forth. The simulation starts with some values of initial sizes of the cities. For example, one can assume $p_i(0) = 1$ for all i. Assuming that $\{\lambda_i\}$ is fixed over time and potentially different across cities, iterating on Equation 6.9 leads to a distribution of population p across cities. In particular, $\{\lambda_i\}$ with uniform distribution leads to the emergence of Zipf's law in size across cities. Probably, the best way to explore this would be to consider a binary interaction model (we will take this up again when we discuss the kinetic exchange model in Section 6.4) in the following form:

$$p_i(t+1) = \lambda_i p_i(t) + \epsilon_t[(1 - \lambda_i)p_i(t) + (1 - \lambda_j)p_j(t)]$$
$$p_j(t+1) = \lambda_j p_j(t) + (1 - \epsilon_t)[(1 - \lambda_i)p_i(t) + (1 - \lambda_j)p_j(t)]. \tag{6.12}$$

It can be shown with a little bit of algebra that for binary interactions (i.e. implementing Equation 6.11 for only two random variables), ϵ would be uniformly distributed. This model was originally proposed in the context of income distribution. We will analyze it in more detail in Section 6.4. It suffices here to note that with λ uniformly distributed across cities, this model generates Zipf's law. While we will not show the derivation of the result here, interested readers can consult Chatterjee and Chakrabarti (2007) which provides mathematical explanations for the origin of Zipf's law from this model.

There are other models that also give rise to Zipf's law in the context of city size. Marsili and Zhang (1998) present an interaction-based dynamic in a non-conservative model that produces Zipf's law. There are models which broadly argue that self-organization in a statistically large system gives rise to the observed behavior (Gan et al., 2006; Batty, 2006). Ghosh et al. (2016) provide yet another approach through agent-based modeling based on the Kolkata paise restaurant framework (see Section 6.6), which also leads to a similar scaling behavior.

However, the empirical validity of Zipf's law in city size distribution has also been questioned in the literature. In the context of US city size distribution, there have been conflicting findings (Black and Henderson, 2003; González-Val, 2010, 2012; González-Val et al., 2014). Eeckhout (2004) showed that lognormal distribution is a good candidate. A more complex picture is presented where the upper tail of US city size distribution seems to be Pareto while the bulk seems to be lognormal (Ioannides and Skouras, 2013; Fazio and Modica, 2015). There are complementary studies of Zipf's law applied to other countries. City size distribution in other countries is highly contested in this regard. There are many studies arguing for the existence of a power law

(Gangopadhyay and Basu, 2009, 2013; Ziqin, 2016) along with many studies that argue the lack of existence of a power law or at least argue in favor of a more nuanced version of the distribution (Anderson and Ge, 2005; Benguigui and Blumenfeld-Lieberthal, 2007; Li and Sui, 2013; Song and Honglin Zhang, 2002). Interestingly, Luckstead and Devadoss (2014) proposed that there may have been a shift in distributional characteristics – the distribution has a better fit with log-normal distribution from 1950 to 1980, while Pareto distribution has a better fit during the census years 1990 and 2010. This is consistent with our observation from Figure 6.2 which also shows that scaling behavior changes over time.

Given the widespread disagreement on the presence of power law and Zipf's law in particular, it is not surprising that many other distributions have been proposed. Some prominent candidate distributions that have been suggested in the literature are lognormal distribution (Eeckhout, 2004), a number of variations around Pareto distributions (see Luckstead and Devadoss, 2017; Ioannides and Skouras, 2013; Reed, 2001; Gómez-Déniz and Calderín-Ojeda, 2015), and q-exponential (Malacarne et al., 2001), among others.

6.4 Inequality and Heterogeneity: Kinetic Exchange Models

So far we have treated interacting agents as effectively zero-intelligence entities. Now we turn our attention to using simple rule-based interactions that can generate macro-level features. Specifically, now we endow the interacting agent with more economic incentives. In terms of methodology, kinetic exchange models have their roots in statistical physics but have found applications in modeling income and wealth distribution.

In a sense, kinetic exchange models depend on the same philosophy as do the spatial interaction models discussed in Section 6.1. The main theoretical underpinning is that macro behavior can be traced back to microscopic rules, but studying microscopic agents in isolation will not lead to an explanation of the macro behavior. The key point is that interaction matters and so does its specific nature.

However, there are two technical differences between kinetic exchange models and spatial interaction models. The first difference is in terms of the system being modeled: kinetic exchange models are used to explain probability density functions of some variables that can be accumulated. The second difference is that the physical interpretation of the *connectivity* between agents does not matter, as opposed to the obvious dependence of the spatial interaction models on the physical nature of interactions (as the name itself suggests).

The main motivation of kinetic exchange models comes essentially from an observed similarity between energy distribution across interacting particles in a gas and income (and wealth) distribution observed in real economies. Historically, the first quantification of the distributional features of such economic variables can be attributed to Vilfredo Pareto (Pareto, 1897), who showed that the top end of wealth distribution displays power law characteristics. Mathematically, a power law distribution of a variable x with coefficient γ is given by

$$f(x) = c.x^{1+\gamma} \quad \text{for } x \geq \bar{x} \tag{6.13}$$

where $\bar{x} > 0$. A more popular version of Pareto's findings can be characterized as the 80–20 rule: that is, 20% of the people possess 80% of wealth. Although his analysis was confined to the population (in a statistical sense) defined by British income tax data, the findings are widely generalizable – not only in terms of income and wealth across time and countries, but also for many other fundamentally different variables including city size distributions (see Section 6.3). The implications are far-reaching. Even though economies differ significantly in terms of the detailed nature of the sociological, political, and even economic nature of their production and consumption processes, there is a *natural* behavior that emerges in income distribution that is independent of the detailed nature of these interactions. A clarification is in order. When we call this emergent feature *natural*, we do not mean that it is fair (from a social welfare perspective). What we imply is that this state, with heterogeneity in outcomes, might be the *most likely* state of the world even in the presence of a wide variety of underlying economic, social, and political rules of interaction (Chatterjee et al., 2007).

However, reliable economic data was very rare in the previous century and exact quantification was extremely difficult. In particular, tax data was reliable only for the very rich. Therefore, data on the bulk of the distribution was essentially missing, and consequently, a large literature was developed around an axiomatic foundation of inequality and wellbeing (Chakravarty, 2009). It is only in the second half of the twentieth century that economists started getting data on income and wealth distribution throughout the population. The main observations that became prominent for explaining the data in terms of statistical models are: (1) the mode of the distribution is at a positive value, (2) the distribution shows positive skewness, and most importantly (3) the bulk of the distribution seems to have a different statistical nature from the right tail (which has a power law feature). The third observation is the most controversial one. Many different classes of distributions have been proposed in the literature that can potentially explain the full income/wealth distribution. The main candidates are lognormal distribution and gamma distribution augmented with a power law tail. In parallel, some studies have focused on purely exponential distributions as well. However, a purely exponential distribution sometimes exhibits good fit with part of the full distribution, but is at odds with the first empirical observation that the mode is not at zero and it clearly cannot produce the power law tail, which is one of the most robust

empirical economic findings ever. Therefore, we will consider only the lognormal
and the gamma distribution with power law tail in what follows.

In terms of statistical descriptions, a lognormal distribution is given by

$$f(x) = \frac{1}{x} \frac{1}{\sigma \sqrt{2\pi}} exp\left(-\frac{(\ln x - \mu)^2}{2\sigma^2}\right) \tag{6.14}$$

where μ and σ are the mean and standard deviation of $\ln x$. A combined definition
of gamma distribution with power law is given by

$$f(x) = \begin{cases} \frac{\beta^\alpha x^{\alpha-1} exp(\beta x)}{\Gamma(\alpha)} & \text{for } x \leq x^* \\ c.x^{1+\gamma} & \text{for } x > x^* \end{cases} \tag{6.15}$$

where $x^* > 0$ and α, β, and c are appropriately chosen constants (see Chapter 2).
The main modeling issue arises with the fact that if the standard deviation σ for
the lognormal distribution is large in magnitude, then it appears linear on a log–
log plot and therefore differentiating from a power law distribution becomes very
difficult. The best way to see it is to take the log of Equation 6.14 to get

$$\ln f(x) = -\ln x - \frac{(\ln x - \mu)^2}{2\sigma^2} - \phi \tag{6.16}$$

where ϕ is a constant. For high σ^2, the above equation would reduce to

$$\ln f(x) = \hat{\phi} - \ln x \tag{6.17}$$

where $\hat{\phi}$ is a constant. Note that this function would be linear on a log–log plot. This
often creates confusion. A standard technique for estimating a power law is to take
log-transformation and use ordinary least square (as described in Section 2.3.5).
While this estimator is biased, many studies used it due to its relative simplicity
and ease of implementation (see Clauset et al. [2009] for a discussion around this
point; they point out statistical problems with this approach and propose a maximum likelihood method with joint determination of cut-off for the right tail). But
for high variance in the data, it may be very difficult to understand whether the
variable really has Pareto distribution (Equation 6.13) or a lognormal distribution
(Equation 6.14). Also, the literature has come up with a lot of instances of variables with supposedly power law distributions which probably are not power laws.
In other words, there are many false positives that the modeler has to be cautious
about.

To give a glimpse into real data, in Figure 6.3 we have plotted the income distribution of about 83,000 in India sampled in 2019 as part of the Consumer Pyramids
Household Surveys (conducted by the Centre for Monitoring Indian Economy: the
same set of households as in Figure 2.6). We also show the fits with gamma and
lognormal distribution (both estimated with maximum likelihood estimator). As
we can see, the gamma distribution does not perform well in capturing the right

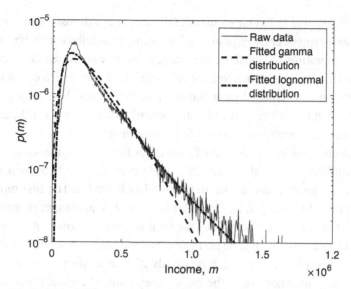

Figure 6.3 Empirical histogram of income distribution among about 83,000 sampled in India in 2019. We show the data in log scale and its fits with gamma distribution and lognormal distribution. Gamma distribution captures the bulk of the distribution, but lognormal distribution captures the tail better. The corresponding consumption distribution is shown in Figure 2.6.

tail. A comparison with Figure 2.6 also shows that consumption inequality is less than income inequality. For consumption, the within-sample standard deviation is 7.2×10^4, whereas for income the same statistic has a value of 2×10^5.

Kinetic exchange models have almost exclusively focused on generating the gamma probability density function and the power law behavior. A significant part of the reason can probably be traced back to the existing result in the statistical physics literature that the energy distribution in a gas is given by gamma distribution. Notably, the physicists Meghnad Saha and Biswambhar Srivastava in their monograph *A treatise on heat* (published in 1931 by Indian Press, Allahabad; their proposal pre-dates almost all the work done by physicists on economic systems) explicitly considered the application of statistical mechanics to model income distribution as a modification of the Maxwell–Boltzmann distribution. In fact, they suggested that the income distribution curve would "will begin with a minimum at 0, rise to a maximum at some point, and thereafter approach the axis of x [i.e. income] meeting it at a great distance" and drew a distribution which resembles a standard gamma distribution (p. 105; reproduced in Chakrabarti [2018]). It is difficult to say if Saha and Srivastava had any access to data at that point of time. The chances are that they came up with the idea purely based on intuition or even speculation. Furthermore, they went on to elaborately conjecture that it would be possible to produce a kinetic theory-based explanation for the emergence of gamma-function-like behavior of income distribution. Almost three decades later, this idea was independently expounded by Benoit Mandelbrot who wrote "(T)here is a great temptation to consider the exchanges of money which occur in economic interaction as analogous to the exchanges of energy which occur in physical shocks between molecules. In the loosest possible terms, both kinds of interactions should lead to similar states of equilibrium" (Mandelbrot, 1960), although no formal treatment or analysis followed for a long time.

A formal analysis in this domain followed in the form of the kinetic exchange model (Chatterjee and Chakrabarti, 2007). Here we present two main variants of the model to generate gamma-like and power law behavior. The first model can be described in the following way. Consider a set S of N agents, each agent holding $m_i(t)$ amount of *money* at time t. Thus the total amount of money in this economy is $M = \sum_j^N M_j(t)$ at any given time point. We assume time is discrete $t = 0, 1, 2, \ldots$. At every time point, two agents are randomly chosen to meet in a marketplace and carry out trade with a portion of the money they hold. The exact process of trade is unimportant. The model simply assumes that due to the trading mechanism there is a random shuffling of money among the agents. Thus one agent randomly gets some ϵ_t fraction of the total money pledged during the trade and the other agent gets the remaining portion: that is, $(1 - \epsilon_t)$ fraction of the total money pledged. This ϵ_t is symmetrically distributed around 1/2. In its basic incarnation, the simplest model assumes that ϵ_t has uniform distribution: that is, $\epsilon_t \sim \text{uniform}[0, 1]$. The last bit of description is about the amount of money retained by the agent. Let us denote that fraction by λ (Chatterjee and Chakrabarti, 2007; Chakraborti and Chakrabarti, 2000; Chakrabarti and Chakrabarti, 2009). The baseline kinetic exchange model can be written as

$$m_i(t + 1) = \lambda m_i(t) + \epsilon_t(1 - \lambda)(m_i(t) + m_j(t))$$
$$m_j(t + 1) = \lambda m_j(t) + (1 - \epsilon_t)(1 - \lambda)(m_i(t) + m_j(t)), \qquad (6.18)$$

where $i, j \in S$ and we are assuming that *savings propensities* are identical across agents: that is, $\lambda_i = \lambda_j = \lambda \in [0, 1]$ for all $i, j \in S$, and $\epsilon_t \sim \text{uniform}[0, 1]$. It can be easily verified that the trading rule is conservative in nature: that is,

$$m_i(t + 1) + m_j(t + 1) = m_i(t) + m_j(t) \qquad (6.19)$$

for all pairs of agents $i, j \in S$ and for all time periods t.

The goal of this model is to find the probability distribution of money m across agents in equilibrium, which turns out to display gamma-distribution-like behavior. To implement this, one first assigns some initial amount of money to the agents $m_i(0)$ and chooses a value of the savings propensity $\lambda \in [0, 1]$ ($\lambda = 1$ would imply that there would not be any non-trivial interactions). At the outset, it is worth noting that the reason for calling the distribution *gamma-like* is that its exact distributional

characteristics are not known yet. A large number of simulation studies and analytical work in the literature has established that the distribution is visually very
similar to a gamma distribution and in fact the first three moments do coincide
with those of gamma distribution. But the moments appear to diverge after that.

The above model cannot produce a power law tail. A very simple modification
allows us to do that. Consider the following trading mechanisms with distributed
savings propensity:

$$m_i(t+1) = \lambda_i m_i(t) + \epsilon_t[(1-\lambda_i)m_i(t) + (1-\lambda_j)m_j(t)]$$
$$m_j(t+1) = \lambda_j m_j(t) + (1-\epsilon_t)[(1-\lambda_i)m_i(t) + (1-\lambda_j)m_j(t)], \qquad (6.20)$$

where $i,j \in S$, $\lambda_i \sim$ uniform[0, 1] and $\epsilon_t \sim$ uniform[0, 1]. This new form of
the savings propensity has two important differences from the earlier one. First,
it is different for different agents and second, the saving propensities are fixed
over time for individual agents. The assumption of saving propensities being
uniformly distributed over a range is important for generating the power law distribution (Chatterjee et al., 2004). A final technical point is in order. We note that λ
being equal to 1 for any agent would lead to extreme concentration of m with that
agent (in every interaction he only gains m and never parts with it). Therefore, we
exclude the parameter λ being exactly equal to 1.

Algorithm 6.3 Kinetic exchange models
Input: Set of agents $\mathcal{N}$; set of savings propensities $\{\lambda_1, \lambda_2, \ldots, \lambda_N\}$; initial asset
vector $\{m_1(0), m_2(0), \ldots, m_N(0)\}$.
Output: Allocation of assets across agents.

---

1. Initiate a set of agents $\mathcal{N}$ with initial asset vector $\{m_1(0), m_2(0), \ldots, m_N(0)\}$.
2. At every time point, choose two agents i and j randomly; they carry out transactions according to Equation 6.18 or 6.20. Repeat this step until each agent
   has participated in at least N number of transactions.
3. Find the probability density function of m.
4. Repeat steps 1–3 above to average over the probability density functions to get
   rid of idiosyncratic fluctuations in the probability density function.

6.5 Dynamics of Languages: Competition and Dominance

Linguistic competition presents an alarming picture of an intrinsically social phenomenon. It is estimated that around 90% of the world's current 6,000 languages
will become extinct very soon. At the same time, about 25% of all languages have
fewer than 1,000 speakers (Crystal, 2002). These observations have led to several
theories about why and how languages become extinct. At the opposite end of

the spectrum are the few languages that dominate the world. Sociologically speaking, there are many reasons for the dominance of a few languages, ranging from historical factors to forced imposition to the trends of globalization, along with geographical factors (Tsunoda, 2006; Amano et al., 2014; Nettle, 1998). In this chapter, we take up this problem in a more mathematical way. Fundamentally, the goal is to model linguistic competition in the way species competition is modeled in evolutionary biology. Thereby, we lose the rich and detailed descriptions of the real world. However, the benefit is that we can perform counterfactual exercises once we get a mathematical model to understand alternative dynamical paths that languages follow.

In the literature, many mathematical models have been used to understand the dynamics of language competition and extinction. These models can be divided into two broad categories. The first category has differential equations describing the evolution of speaker concentration or fraction of the population who speak certain languages (Kandler and Unger, 2018). The second category is comprised of agent-based models where individuals represent nodes in a network defining linguistic interactions (Zhang and Gong, 2013). In the present context, we solely focus on a few important models in the first category.

Before describing the models, we will describe some real-world data on language dynamics. In Figure 6.4, we show the evolution of the fraction of native speakers speaking Indian languages over a period of 30 years. The data was collected from the decadal Indian census reports of 1991, 2001, and 2011. While we might except the dynamics of language evolution to be slow, even in this 30-year period we see trends in the fraction of speakers speaking these languages. Hindi, for example, shows a slight upward trend and other languages show slightly downward trend. Although we cannot infer much based on only three observations in the time domain, this provides a starting point to think about how different languages compete and evolve.

6.5.1 The Monolingual Model

One of the simplest monolingual models was proposed by Abrams and Strogatz (2003). This is a microscopic model for competition between two languages, similar to the prey–predator Lotka–Volterra model. However, the original Lotka–Volterra model represented one predator and one target, whereas this model considers both the languages to behave like prey and predator to one another, resulting in the irreversible deaths of languages.

Let us consider the proportion of monolingual speakers of languages X and Y as N_X and N_Y, respectively, with the normalization that $N_X + N_Y = 1$. By definition, a monolingual speaker speaks only one of the two languages. The rate at which N_X and N_Y change over time is given by the following:

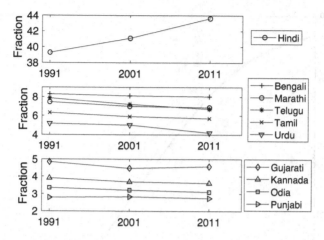

Figure 6.4 Evolution of the fraction of native speakers of various Indian languages from 1991 to 2011 at three time points separated by decades – 1991, 2001, and 2011.

$$\frac{dN_X}{dt} = N_Y P_{YX} - N_x P_{XY} \tag{6.21}$$

where P is the instantaneous probability by which a given individual switches language, that is, switches from X to Y or vice versa. This probability is defined by the following expression:

$$P_{YX} = s N_X^a$$
$$P_{XY} = (1 - s)(1 - N_X)^a, \tag{6.22}$$

where a and s are parameters. One can show that the stable equilibria of the dynamics occur at $N_X = 0$ or $N_X = 1$ for $a > 1$. Therefore, irrespective of the initial condition, the system will end up on either of the equilibrium points with the complete extinction of one language. We show some simulation results to visualize the transitional dynamics, using identical initial conditions $N_X(t = 0) = 1/2$ for both the languages. Figure 6.5 shows the evolution of the fraction of speakers for different values of the parameter a around 1.31, which Abrams and Strogatz found to be the average estimate across various cultures. The parameter s captures the prestige of a language, and s being less than 1/2 represents a scenario where X has lower prestige than Y. Intuitively, Abrams and Strogatz argued that this parameter is a *measure of threat* to a given language.

6.5.2 Models with Bilinguals

Minett and Wang (2008) extended the above model to consider bilingual agents, which includes the possible coexistence of languages. In this model, there are three

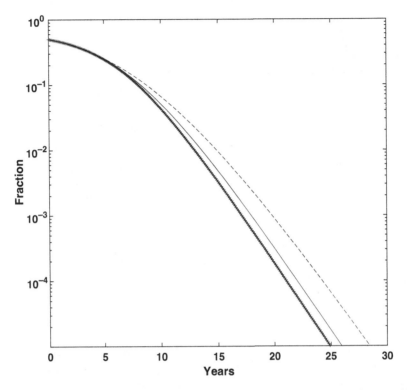

Figure 6.5 Fraction of speakers of language X from Equations 6.21 and 6.22 with parameter values $a = 1.4$, 1.31, and 1.2 (from left to right), at $s = 0.4$.

types of agents: speakers who speak only language X, speakers who speak only language Y, and speakers who speak both X and Y. We denote the third kind by Z. Let us denote the number of speakers in three languages by N_X, N_Y, and N_Z, respectively. Let us assume that a direct transition from language X to Y or vice versa is not possible. Therefore, only four transitions are possible: $X \to Z$, $Z \to X$, $Y \to Z$, and $Z \to X$. The model is given by

$$\frac{dN_X}{dt} = k_{ZX}N_Z - k_{XZ}N_X$$
$$\frac{dN_Y}{dt} = k_{ZY}N_Z - k_{YZ}N_Y \qquad (6.23)$$

where k_{XZ}, k_{ZX}, k_{YZ}, and k_{ZY} are the rates of linguistic transitions that depend upon various real-life factors (e.g. mortality rate, language status, and so on). Given the normalization $N_X + N_Y + N_Z = 1$, the above equations can be rewritten as

$$\frac{dN_X}{dt} = s(1 - N_X - N_Y)(1 - N_Y)^a - (1 - s)N_X N_Y^a$$

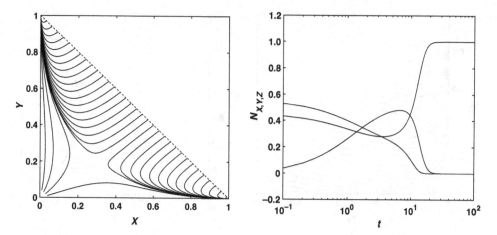

Figure 6.6 Example of direction field corresponding to the Minett–Wang model (left-hand panel). Evolution of the system is shown with the initial conditions $N_X(t = 0) = 0.55$, $N_Y(t = 0) = 0.45$, $N_Z(t = 0) = 0$ (right-hand panel).

$$\frac{dN_Y}{dt} = (1 - s)(1 - N_X - N_Y)(1 - N_X)^a - sN_Y N_X^a. \tag{6.24}$$

This system of equations has three equilibrium points in N_X, N_Y, and N_Z. Out of these, two points (1, 0, 0) and (0, 1, 0) correspond to all the speakers only being monolingual in X or monolingual in Y. These are stable for $a > 1$. The third equilibrium admits agents who are bilingual, but it is unstable. However, for some values of $a < 1$, the stability of the system flips. Generally, it is the third equilibrium which is of interest as that allows for the possibilities of bilinguals to exist. We show an example of the direction fields and the evolution of the model in the Figure 6.6.

The above model assumes that the rate by which the monolingual $X(Y)$ becomes bilingual is proportional to the number of monolingual speakers of other languages $N_Y(N_X)$. Heinsalu et al. (2014) proposed a modification in which they assumed that the rate of change at which the monolingual $X(Y)$ become bilingual would be proportional to the total number of speakers of language $Y(X)$ (which also includes bilinguals) – specifically, to the sum $N_Y + N_Z(N_X + N_Z)$. Like the earlier model, this model also has multiple equilibrium points with a stable equilibrium where bilingual speakers exist. We show an example of the direction fields and the evolution of the fraction of speakers in Figure 6.7.

6.6 Emergence of Coordination and Anti-coordination: Bounded Rationality and Repeated Interactions

One of most important differences between economic systems and physical systems is that economic agents are not passive particles; they learn, adapt, and

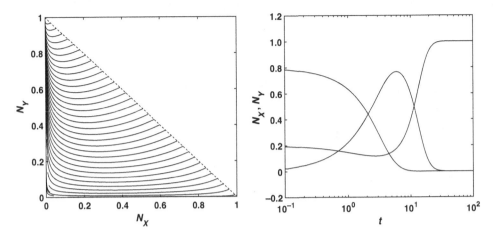

Figure 6.7 Example of direction fields corresponding to the Patriarca–Heinsalu model (left-hand panel). Evolution of the system is shown with the initial conditions $N_X(t = 0) = 0.8$, $N_Y(t = 0) = 0.2$, $N_Z(t = 0) = 0$ (right-hand panel).

strategize. In this section, we introduce a stylized model of an anti-coordination game with multiple agents. This game is rich enough to introduce strategies and different learning schemes. But before getting to the main problem statement, we will take a detour through some game theoretic terminology. That will help us to fix some ideas about the solution of the game.

A Brief Description of Strategic Games

A strategic game is defined by a triplet $\langle N, A, \pi \rangle$, where N denotes a set of players, A_i denotes a set of actions or strategies for the ith player in N, and π denotes a payoff function which maps action profiles into real number, that is, $\pi : A \to \mathbb{R}$. A game captures a scenario where multiple players engage in strategic interactions, attempting to maximize their own payoffs. Note that maximizing the payoff here is a complicated task. To fix the notion, let us assume that there are only two players A and B, each of whom has a fixed number of strategies. If player B did not have a choice over their strategies, then player A would be able to maximize their own payoff quite easily. They could simply play the strategy for which their payoff is the highest. Similarly for player B, if A did not have any choice over strategies then maximizing B's payoff would be pretty easy. The non-trivial feature of strategic games arises when these individual payoff-maximizing strategies do not coincide. Thus it is quite possible that if both of the players play their own payoff-maximizing strategies without paying attention to what the opponent is playing, then both might end up in a worse scenario than other better and feasible outcomes. Therefore, the strategic interaction becomes *non-trivial* when each of the agents takes the other agent's possible decisions into their own decision-making processes.

Game theory itself has a huge literature with wide-ranging applications in economics, political science, sociology, evolutionary biology, and so forth. Interested readers should consult standard textbooks on game theory for in-depth discussion on the equilibrium concepts in games, for example, Osborne (2004). For an interpretation of economics in the context of evolution, readers should consult Gintis (2009). At this point, it would also be useful to note that bounded rationality has attracted the attention of economists for a very long time (Simon, 1997). Herbert Simon made an early and enormous contribution to the development of the idea of how computation in the human mind leads to rationality (see e.g. Simon, 1957). In more recent times, many new developments have taken place in mathematical models of cognition in the context of decision-making processes – sparse matrix operator (Gabaix, 2014), rational inattention (Sims, 2003), and reinforcement learning (Nowé et al., 2012), among others.

In what follows, we will very briefly touch upon the game theoretic solution of the proposed game. Most of our focus will be on statistics-based and simulation-based approximate analysis.

6.6.1 The Kolkata Paise Restaurant Problem: A Baseline Resource Allocation Game

The Kolkata paise restaurant (KPR) problem is an N-agent, M-choices problem where agents simultaneously strategize to avoid making the same choice. To understand the nature of interactions in this game, it is better to see it as a competition for resources rather than allocation of resources. To fix the terminology for the baseline version of the game, let us imagine that choice represents a restaurant and each restaurant can accommodate only one person for lunch. Thus there are N agents strategizing about where to get their lunch from N choices of restaurants. This game assumes that time is discrete, $t = 0, 1, 2, \ldots$, and at every time point (which might represent a day), each agent can make only one attempt to secure a spot in a restaurant. Note that by assumption each restaurant has only one spot to offer and they do not treat customers preferentially. Thus if more than one agent chooses the same restaurant, then each of them is equally likely to get the lunch and only one of them randomly gets it (the resource, i.e. *lunch*, is non-divisible). This leads to congestion. Herein lies the non-trivial strategic decision-making part of this game. It is in the interest of each agent to avoid all other agents arriving at restaurants. The question is, if they cannot communicate with each other, how would they coordinate to arrive at a *socially efficient* outcome where there is as little congestion as possible? In the original work that proposed the problem (Chakrabarti et al., 2009; see also Ghosh and Chakrabarti, 2017; Chakrabarti et al. 2017), the authors added another level of complexity in the game by introducing rank-ordering among the restaurants in terms of quality of service.

We will introduce some formal notations to first describe the theoretical solution in the form of Nash equilibria. The first thing to note is that we will analyze it as a one-shot game. Although it can be thought of as a repeated game following traditional game theory, we will maintain the assumption of a one-shot game for simplicity. Also, this will allow us to delve deeper into the literature on inductive reasoning in a simpler and more intuitive way. Then we will proceed toward an agent-based approach to see how close to the theoretical solution we can get depending on a simple rule of thumb. It is useful to note here that although the Nash equilibria in this game are conceptually easy to find, it is highly unlikely that in a real-world game like this people would be able to coordinate to approach the equilibrium solution. The coordination cost may be too high. If the agents do not have global information (in which case the cost of information gathering and processing may become too high), it is difficult to sustain the equilibrium, especially when the restaurants can be rank-ordered with high dispersion in perceived utility. However, simple rule-of-thumb behavior can take the system close to such equilibria. Let us elaborate on the idea below.

The set of agents is denoted by $\mathcal{N} = \{1, 2, \ldots, N\}$. The rank-ordering of restaurants can be represented by a *utility* vector $u = \{u_1, u_2, \ldots, u_M\} \in \mathbb{R}_+^M$. Throughout, we will maintain the assumption that $M = N$. For simplicity, we can assume that all agents agree on the same rank-ordering $0 < u_N < u_{N-1} < \cdots < u_1$. The ith agent has a strategy set at time t that can be written as $s_{it} \in \{1, 2, \ldots, M\}$ where the strategy at every point is to choose a single restaurant at time t. While this principle is generalizable to mixed strategies (for a textbook treatment of mixed strategies, see Osborne [2004]), here we stick to deterministic strategies to keep the exposition simple. The corresponding payoffs are given by $\pi_t = \{\pi_{1t}, \pi_{2t}, \ldots, \pi_{Nt}\}$. Note that if the jth agent chooses restaurant k and no one else contests that choice, then $\pi_j = u_k$. However, if two or more agents (say n) choose the same restaurant k, then the expected payoff of all those agents is u_k/n as all of them are equally likely to get it. There is an easy way to represent the combined payoffs. Given an action profile $S_t \in S^N$, the payoff of the ith player is $\pi_{it} = u_{s_i}/n_i(S_t)$, where $n_i(S_t)$ is the number of agents at time t playing strategy profile S_t, having made the same choice as player i.

The intuition for finding a pure strategy Nash equilibrium is simple. There has to be one-to-one matching between the set of agents and the set of restaurants and the agents should not deviate from this outcome. The way to stop deviation is obvious. Note that if each agent gets access to a unique restaurant, then the payoff of the *i*th agent is simply $\pi_i = u_{s_i}$. Therefore, there will be one agent j with the lowest utility $\pi_j = u_{s_j} = u_N$. This agent has the highest incentive to deviate from this configuration, and the best he can do is to try going to the best restaurant. But given that there already one agent allocated there, the payoff from deviation would be $\pi_j = u_1/2$. Thus if $u_N > u_1/2$, then even the agent with the worst outcome would not want to deviate. Therefore, we can guarantee the existence of a pure strategy Nash equilibrium under the condition $u_N > u_1/2$.

A related concept we would like to explore is *social efficiency*. Note that in the case $M = N$, any vacancy is necessarily related to wastage of a lunch. Thus congestion in this model is not only bad for the agents, but also bad from the *production* point of view. In this case, we will say that a strategy profile $S_t = \{s_1, s_2, \ldots, s_N\}$ is socially efficient (also known as *Pareto* efficient; the concept is named after Vilfredo Pareto, the same scientist after whom Pareto distribution is named, but there is no connections between the two concepts) if there is no congestion: that is, $s_i \neq s_j$ for all pairs (i and j) of players. In this case, it is easy to see that all pure strategy Nash equilibria are socially efficient. This relationship holds in the opposite direction as well: that all socially efficient outcomes are Nash equilibrium. We will not give the proof here; see Chakrabarti et al. (2009) for a proof.

6.6.2 Inductive Reasoning and Bounded Rationality

Although we can define the socially efficient outcome and the Nash equilibria easily, it should be evident by now that these are fundamentally static concepts and we have not explained how independent agents might achieve such coordination. Obviously, we are ruling out the possibility of a dictator (or social planner) who dictates an allocation and whom agents simply obey; that does not really address the problem of how *autonomous* agents can coordinate and, more importantly, how such attempts for coordination might fail, especially when the number of competing agents is very large.

We posit that it is more natural and computationally far less burdensome for real-world agents to utilize rule-of-thumb behavior for playing such multi-agent games with imperfect information. This idea can be traced back to the El Farol Bar problem proposed by W. Brian Arthur (Arthur, 1994). In fact, the heading of this section is also taken from the name of the paper where Arthur proposed this game.

The main idea is that given the large combination of possible strategies that the agents might play, leading to a commensurate number of potential outcomes, it might be easier to do away with the ideals of *complete knowledge* and *rationality*. In fact, we go to the opposite extreme of allowing players to possess very limited amount of knowledge and memory, and we endow them with simple stochastic strategies to play the game. The goal is to see how close to the Nash equilibria the agents can get using such simple behavioral rules of interaction.

Below, we describe two strategies for which analytical solutions exist (Chakrabarti et al., 2009). Many other behavioral strategies have been studied in the literature with varied success rates. Readers are encouraged to try out alternative strategies in their own simulations of this game.

No learning rule: The simplest learning rule is that of "no learning." The easiest way to operationalize it is to imagine a behavioral rule which simply assigns $1/M$ probability to each restaurant, which does not respond to past behavior, success, or failure. This is a useful benchmark to start from. While this behavioral rule can be imposed and a model game can be easily simulated, one can utilize a very useful but simple trick to theoretically predict the resultant utilization fraction.

Let us denote the number of agents in any arbitrary restaurant, say the kth one, by m_k. We want to find out the distribution of m_k. For ease of exposition, let us denote N/M by λ. Now, if each agent randomly chooses a restaurant, the probability that they will choose the kth restaurant is $p = \frac{1}{N}$. Thus for any given restaurant, the probability that m agents will select it is given by the probability that exactly m agents select that restaurant and $N - m$ do not. Therefore, this probability can be written as

$$P(m) = \binom{N}{m} p^m (1 - p)^{N-m}. \tag{6.25}$$

This can be recognized as a Poisson distribution. By substituting n and M and letting the number of players be large, we get

$$P(m) = \binom{\lambda M}{m} \left(\frac{1}{M}\right)^m \left(1 - \frac{1}{M}\right)^{\lambda M - m}. \tag{6.26}$$

In the limit ($M \to \infty$), the distribution converges to the following:

$$P(m) = \frac{\lambda^m}{m} exp(-\lambda). \tag{6.27}$$

For the symmetric case, we have $N = M$ implying $\lambda = 1$. Therefore, the fraction of people unable to find a match would be $P(m = 0) = exp(-1)$ upon substitution of the value of λ. Thus, the fraction of restaurants occupied would be $P(m \neq 0) = 1 - exp(-1) \approx 0.63$.

Minimal learning rule with limited memory: There are many different ways one can introduce complex decision-making rules in this game. Here we analyze a very simple modification of the above **No learning** rule. The idea is that if an agent on any given day secures a match, they go to the best restaurant the next day. The remaining agents pursue a rule of no learning: that is, they choose randomly. This strategy introduces two important features that have been studied in the literature in great detail. There is now an effect of memory (since this rule is now conditional on past success), and there is learning, although in a very primitive and ad hoc form.

There is an easy derivation of the resulting utilization fraction. Let us denote the number of agents who could get lunch on day t by $f_t N$. Note that the next day they all converge on the best-ranking restaurant, but only one of them will be able to finally secure a match. The remaining $N - f_t N$ agents will choose randomly out of N restaurants. Therefore, we can ignore the successful agents at time t (when $N \to \infty$) and simply follow the method proposed in the **No learning rule** above to get a recursive solution for the fraction f (defining the ratio of the number of agents randomly choosing with respect to the number of available restaurants by λ_t):

$$f_{t+1} = 1 - exp(-\lambda_t)$$
$$= 1 - exp(-(1 - f_t)). \tag{6.28}$$

It can be checked that the fixed point f for the expression

$$f = 1 - exp(f - 1) \tag{6.29}$$

is given by $f^* \approx 0.43$.

There are two major takeaways from these simple exercises. First, the **No learning rule** is symmetric in nature and every period about 63% of agents can secure a match. Second, learning or responding to past success could lead to worse outcomes. Therefore, not all learning algorithms are necessarily beneficial (Chakrabarti and Ghosh, 2019; Chakraborti et al., 2015).

We provide here a pseudo-code for simulating the model. Many new learning rules can be endowed on the agents and run through the model to see the efficacy of the rules. One interesting exercise would be to endow different agents with different rules and then see which one dominates the others. On a final note, the extant literature on this game considered the time required for convergence as an important metric for evaluating the performance of different learning rules. We have instead focused only on the allocative and matching perspectives for the description of the model, following a more economics-oriented approach. In many real-life applications, such as like centralized algorithms to match customers to cabs (see Chakrabarti et al. [2017] for a review), time requirement would indeed be an important factor to consider.

6.7 Realism vs. Generalizability

Let us now summarize the above discussion. We have worked through a series of progressively more complex models that allow us to connect microscopic interactions to macroscopic behavior. Here we take a pause and consider the trade-offs

---

Algorithm 6.4 Kolkata paise restaurant problem

Input: Set of agents $\mathcal{N}$; set of strategies for agents in $\mathcal{N}$; set of restaurants $\mathcal{M}$.

Output: Allocation of agents across restaurants and realized payoffs.

---

1. Initiate a set of agents $\mathcal{N}$ and a set of restaurants $\mathcal{M}$.
2. All agents are endowed with strategy $\{S_i\}_{i \in \mathcal{N}}$. This strategy can be stochastic in nature and may utilize local information (e.g. past success rate) or global information (e.g. the common ranking of restaurants), and can be memory-dependent.
3. At every time point, each agent chooses one restaurant based on their information set and the strategy they are endowed with.
4. Match the agents to their choices of restaurants. If a tie arises due to multiple agents choosing the same restaurant, then only one agent is chosen randomly to break the tie. The rest of the agents do not get a match in that period. Similarly, unmatched restaurants remain vacant in that period.
5. Calculate the rate of utilization as the ratio of the number of matched restaurants to the total number of restaurants (or the fraction of agents who could secure a match).
6. Repeat steps 3–5 to generate a time series depicting the evolution of the rate of utilization.

---

that one has to consider in using the models to capture specific mechanisms or behavior.

In all the above cases, the models discussed are deliberately stylized in nature. There are pros and cons of this. The obvious downside of such models is that they are not particularly realistic and fail to map the mechanism driving the results in the model on to the exact phenomena in reality. However, we consider this to be more of a virtue than a vice. The basic proposition of complex systems science is that many economic, social, and natural phenomena are different at the system level, but they still share a fundamental core similarity that drives their dynamics. Therefore, these models are often by choice constructed as the simplest possible mechanisms that are able to capture the targeted behavior, as opposed to making them realistic and losing generalizability.

This also shows the usefulness *and* limitations of the present approach. If a very specific intervention is necessary (e.g. taxation in the Indian economy to raise tax revenue by a specific percentage), then a kinetic exchange model would be a bad choice for policy advice since it clearly does not incorporate important strategic decision-making counterparts in the model, and it lacks depth in description of how taxes impact labor income vis-à-vis income from capital stock. However, if one asks what kind of minimal ingredient model can explain the differences in

Gini coefficients across developed and developing countries, then kinetic exchange models would be quite informative and useful, due to ease of calibration and simulation. Thus eventually it depends on the modeler to decide what kind of trade-offs, in terms of realism versus generalizability, need to be made in order to answer a given question.

Each of these topics (ranging from social segregation to city size dynamics to competition among languages) has its own nuances and details, which cannot be reduced to a set of physical or social laws. As Philip Anderson famously stated "Psychology is not applied biology, nor is biology applied chemistry" (Anderson, 1972). While it is not possible to do justice to the rich literature behind all the models described above, here we take up the case of inequality to indicate why and how history and institutions matter.

Inequality has been at the forefront of political economy, and even economics in general, from the time of the inception of the subject in any form. The recognition that income and wealth are not equally distributed, and the corresponding philosophical and policy-oriented implications, are there in all of the major works in economics. The distributional features are only one part of the whole picture. There are other quantifiable features, such as inequality measures, that are independent of algebraic forms of the distributional characteristics (see Chakravarty, 2009, for example). Furthermore, the sources of income and wealth are also important factors for studying distributional features as well as for policy-making. In what follows, we will describe some of the *mechanisms* and sources of inequality. One reason for discussing them is to see whether a process as complex as economic inequality can really be understood in terms of simplistic toy models. Another reason is to see if there are ways to modify the models further in order to discover more realistic mechanisms.

In recent times, the discussion of inequality has gone through a revival in the academic as well as public domain in the aftermath of the 2008 financial crisis. Thomas Piketty, in his widely popular book *Capital in the Twenty-First Century*, documented the evolution of inequality in terms of shares of income and wealth across deciles in the whole distribution over time and across regions (Piketty, 2014). One of his key findings is that many countries, including the USA, are experiencing increasing trends in income inequality. In this respect, it is useful to differentiate between labor income and capital income as is commonly recognized in the economics literature. Labor income denotes the income that people earn for providing their labor, whereas capital income is the income generated by assets held by people. Typically it is the case that the right tail of the income distribution is dominated by return from capital and the process that generates inequality in return from capital is fundamentally different from the process generating labor income. For example, labor income would be potentially dependent on the quality of labor, whereas capital income would be driven by risk and return in the market. Potentially the rich-get-richer mechanism can be active in both channels of income. Rich parents' children would get access to more social and human capital and therefore would be more likely to earn more themselves. Similarly, people with large shares of capital can invest in more profitable investment projects earning higher rates of return. But this phenomenon is likely to be more prominent in the case of capital earnings, at least in the short run as opposed to long-run generational changes. Much of the present-day discussion and debate revolves around the share of income between labor earnings and capital earnings, where the share of labor seems to be almost monotonically decreasing in many countries (Karabarbounis and Neiman, 2014).

The above discussion elucidates some major lacunae, in terms of social and economic factors, in kinetic exchange models. In this class of model, the mechanism of inequality-generating processes depends on exogenously specified distribution of the random coefficient ϵ and the distribution of the savings propensity λ. These models clearly fall short of modeling decision-making processes or the actual economy in a credible way. There are some works (Chakrabarti and Chakrabarti, 2009; Quevedo and Quimbay, 2020) that have attempted to forge a link between more realistic parametric description of trading processes in kinetic exchange models, and the resultant behavior of the economy. In particular, the first paper shows that a formal model can be written with utility-maximizing agents that engage in trading and the resultant behavior can be exactly mapped into Equation 6.18. However, much more work in this domain is needed before a stylized model will be rich enough to explain cross-sectional distributional features as well as the time series behavior of inequality (Atkinson et al., 2011).

But there are some positive aspects of this kind of modeling as well. There are very well-defined economic models based on explicit dynamic optimization that can explain the inequality dynamics and some of the distributional features quite accurately (Benhabib et al., 2011). However, these models often have a hard time explaining the distributional features in a parametric setup. What we mean by that is as follows – these models perform well to explain shares of income/wealth accruing to different percentiles of the population, but they do not really explain well the exact parametric distributions such as whether the distribution is gamma or lognormal and so forth. In such cases, kinetic exchange models have a unique usefulness in terms of simplicity and ease of mapping to known algebraic forms of distributions (Chatterjee and Chakrabarti, 2007).

In the end, we note a fact that all scientific disciplines recognize: all models are wrong, but some are useful. We note that even among the useful models, different classes of models are useful for different purposes. Therefore, abstract models from complex systems literature can coexist with more intricate system-dependent models with gains from symbiosis. There is no particular reason that they have to be substitutes rather than complements. Finally, the philosophy of complex systems has attracted a lot of well-deserved attention in the literature. We have followed the approach that usefulness of the models is the most important criteria for us. In that way, we are in sync with Esther Duflo, a Nobel Prize winning economist, when she mentioned that economists should be more like plumbers (Duflo, 2017). The idea is that a literature can evolve around a set of toolkits which should be large enough to capture the mechanisms of wide variety, should be flexible enough to be subject to tinkering as need be, and finally, should be useful enough to work in the real world. After all, just like economics, most of the current complex systems paradigm represents inexact science. It would also be useful to note here that Duflo, along with Avijit Banerjee and Michael Kremer, won their Nobel Prize precisely for working on identification problem in causal inference, specifically on *randomized controlled trials* which also adaptively build on econometric theory along with abstract models. There have been enthusiastic propositions in favor of data making models redundant (Anderson, 2008), but it also seems that a blind adherence to data-crunching may not provide a very successful paradigm (Jackson, 2019). As we have emphasized, both are important, and a data-driven approach complements theory in a big way.

Epilogue

In this book, we have viewed complex systems through the lens of a mix of statistics, econometrics, network theory, and machine learning, emphasizing the empirical aspects throughout. The complex systems literature had some initial bursts of excitement from the realization that there can be universalities in the behavior of systems, which loosely speaking translates into an idea that a detailed description of a system may not be consequential for the observed emergent phenomena. For example, there have been some proposals that maybe the archetypal complex system model in the form of sand-piles (Bak, 2013) would provide explanations of the appearance of power laws in wildly different contexts ranging from geophysics to brain science to economics (Chakrabarti and Marjit, 1995). There have been similar conjectures in the past about other systems as well, which displayed some potential for unifying theories (e.g. Chakrabarti and Chakrabarti [2009] showed that entropy maximization in particle interactions and utility maximization in economic interactions have a deep similarity). However, the success of such models in terms of falsification with respect to the empirical data remains weak at best and often the scope is limited. So where do we go from here? In order to think about this question, it will be useful to take a detour in terms of the lineage of the way complexity theory is currently interpreted.

In much of the book, we have emphasized economic contexts, scenarios, and interpretations. One may find this surprising given that the physics literature has a broad repertoire of complex systems and therefore might make a more natural point of reference. In fact, the 2021 Nobel Prize in Physics was awarded to Giorgio Parisi (along with Syukuro Manabe and Klaus Hasselmann) recognizing his contribution to the understanding of physical aspects of complex systems and the "interplay of disorder and fluctuations in physical systems from atomic to planetary scales" (quoted from the prize citation). However, there is a specific reason behind our approach. Economics has a major stake in this literature. Historically, Adam Smith has been recognized as the father of economics. Interestingly, a slightly unorthodox interpretation of his famous "invisible hand" paradigm for markets provided probably the first recognizable description of self-organization in a multi-agent system arising out of non-trivial interactions. This paradigm pre-dates the famous "Game of Life" by almost 200 years. The mathematician John Conway came up with this model in 1970. Starting around the same time, Robert May's influential work on simple dynamics leading to chaotic behavior and analysis of (in-)stability of interacting systems shed light on the relationship between stability and emergence. Although May's focus was on ecological systems, these works found important applications in economic and financial systems, as we discussed in Chapter 1. Benoit Mandelbrot developed fractal geometry and multiple types of time series toolkits which are widely applicable across different domains of complex systems. We reviewed some of them in Chapter 3; interestingly, Mandelbrot also thought about models of economic inequality in terms of kinetic exchange models, as we discussed in Chapter 6. Additionally, W. Brian Arthur provided an impetus to the growth of complexity science in economics, specifically in terms of path dependence of technological evolution and establishing that modes of interactions matter for aggregate market behavior. In particular, his El Farol Bar problem has been an important benchmark model for at least two streams of the literature. First, it led to *minority games* which provided a useful paradigm in modeling financial complexity (Challet et al., 2004). Second, its generalization in the form of the Kolkata paise restaurant problem (discussed in Chapter 6) expanded its scope to incorporate a more complete game theoretic formulation of agent-based models.

A parallel literature recognized that complexity begets complexity. In a combinatorial sense, a heterogeneous system, if prone to recombinations among the existing constituent entities, will lead to more heterogeneity simply by small but steady steps of combinatorial processes (Kauffman, 2000). In fact, such a system will show glacially slow increase in heterogeneity for a long time, before exhibiting explosive growth giving rise to a "hockey-stick" pattern. This was noted by Stuart Kauffman and this idea has been subsequently applied in the context of the growth of technology that the human society has experienced over the entire history (Koppl et al., 2021). Therefore, the timescale also matters to define the system and the *degree* of complexity embedded in it. This in turn indicates that a theory which may capture a given kind of complex behavior maybe insufficient at a bigger or a smaller scale. Different types of behavior may emerge at different scales, both in terms of time and structure.

In a sense, the lack of one single unifying framework may seem to be problematic for the complexity theory that is purportedly general enough to cut across disciplinary boundaries. In fact, Herbert Simon stated that "systems of such diverse kinds could hardly be expected to have any nontrivial properties in common" (quoted from Simon [1991]). Even the famous power law behavior has been contested in terms of its value addition to scientific understanding of underlying mechanisms (Stumpf and Porter, 2012). However, we have a slightly different point of view here. Possibly the search for stable and robust statistical laws like scaling behavior in all social and economic systems is a futile endeavor (at least, as of now). It's quite likely that patterns do exist in the socio-economic world but that they are short-lived and non-robust by the standards of physical systems. For example, models of language dynamics can display an equilibrium behavior where a certain language dies and another competing language thrives. The transitional dynamics may not be particularly interesting from a mathematical point of view. However, at the timescale of human lives, probably transitional dynamics is the most intriguing area here as such dynamics may lead to social conflict and the collapse of cooperation, and may even impact economic and cultural growth. If we zoom in on each of these phenomena, potentially we will uncover the defining features of complex systems: heterogeneity, interactions, and emergence. Thus, complexity theory may evolve in future into a collection of toolkits to study complex behavior in a wide array of dissimilar contexts, rather than a predetermined set of theories. The present work is a step in that direction.

References

Abe, Sumiyoshi, and Suzuki, Norikazu. 2004. Scale-free network of earthquakes. *EPL (Europhysics Letters)*, **65**(4), 581–586.

Abrams, Daniel M, and Strogatz, Steven H. 2003. Modelling the dynamics of language death. *Nature*, **424**(6951), 900.

Acemoğlu, Daron, Carvalho, Vasco M, Ozdaglar, Asuman, and Tahbaz-Salehi, Alireza. 2012. The network origins of aggregate fluctuations. *Econometrica*, **80**(5), 1977–2016.

Acemoğlu, Daron, Ozdaglar, Asuman, and Tahbaz-Salehi, Alireza. 2015. Systemic risk and stability in financial networks. *American Economic Review*, **105**(2), 564–608.

Acemoğlu, Daron, Akcigit, Ufuk, and Kerr, William R. 2016a. Innovation network. *Proceedings of the National Academy of Sciences*, **113**(41), 11483–11488.

Acemoğlu, Daron, Akcigit, Ufuk, and Kerr, William R. 2016b. Networks and the macro-economy: An empirical exploration. *NBER Macroeconomics Annual*, **30**(1), 273–335.

Aggarwal, Mayank, Chakrabarti, Anindya S, and Dev, Pritha. 2020. Breaking "bad" links: Impact of Companies Act 2013 on the Indian Corporate Network. *Social Networks*, **62**, 12–23.

Aït-Sahalia, Yacine, and Jacod, Jean. 2014. *High-frequency financial econometrics*. Princeton University Press.

Albert, Réka, Albert, István, and Nakarado, Gary L. 2004. Structural vulnerability of the North American power grid. *Physical Review E*, **69**(2), 025103.

Alpaydin, Ethem. 2020. *Introduction to machine learning*. MIT Press.

Altman, Naomi, and Krzywinski, Martin. 2015. Association, correlation and causation. *Nature Methods*, **12**(10), 899–900.

Amano, Tatsuya, Sandel, Brody, Eager, Heidi, et al. 2014. Global distribution and drivers of language extinction risk. *Proceedings of the Royal Society B: Biological Sciences*, **281**(1793), 20141574.

Amaral, Luís A Nunes, Scala, Antonio, Barthelemy, Marc, and Stanley, H Eugene. 2000. Classes of small-world networks. *Proceedings of the National Academy of Sciences*, **97**(21), 11149–11152.

Amari, Shun-ichi, and Wu, Si. 1999. Improving support vector machine classifiers by modifying kernel functions. *Neural Networks*, **12**(6), 783–789.

Andersen, Torben, Bollerslev, Tim, and Hadi, Ali. 2014. *ARCH and GARCH models*. John Wiley & Sons.

Anderson, Chris. 2008. The end of theory: The data deluge makes the scientific method obsolete. *Wired Magazine*, **16**(7), 1–2.

Anderson, Gordon, and Ge, Ying. 2005. The size distribution of Chinese cities. *Regional Science and Urban Economics*, **35**(6), 756–776.

Anderson, Philip W. 1972. More is different. *Science*, **177**(4047), 393–396.

Angrist, Joshua D, and Pischke, Jörn-Steffen. 2008. *Mostly harmless econometrics*. Princeton University Press.

Antal, Tibor, Krapivsky, Paul L, and Redner, Sidney. 2006. Social balance on networks: The dynamics of friendship and enmity. *Physica D: Nonlinear Phenomena*, **224**(1–2), 130–136.

Arthur, W Brian. 1994. Inductive reasoning and bounded rationality. *American Economic Review*, **84**(2), 406–411.

Athey, Susan, and Imbens, Guido W. 2019. Machine learning methods that economists should know about. *Annual Review of Economics*, **11**, 685–725.

Atkinson, Anthony B, Piketty, Thomas, and Saez, Emmanuel. 2011. Top incomes in the long run of history. *Journal of Economic Literature*, **49**(1), 3–71.

Axtell, Robert L. 2001. Zipf distribution of US firm sizes. *Science*, **293.5536**, 1818–1820.

Babcock, KL, and Westervelt, RM. 1990. Avalanches and self-organization in cellular magnetic-domain patterns. *Physical Review Letters*, **64**(18), 2168–2171.

Backstrom, Lars, Boldi, Paolo, Rosa, Marco, Ugander, Johan, and Vigna, Sebastiano. 2012. Four degrees of separation. Pages 33–42 of: *Proceedings of the 4th Annual ACM Web Science Conference*.

Bailey, Michael, Cao, Rachel, Kuchler, Theresa, Stroebel, Johannes, and Wong, Arlene. 2018. Social connectedness: Measurement, determinants, and effects. *Journal of Economic Perspectives*, **32**(3), 259–280.

Bailey, Michael, Farrell, Patrick, Kuchler, Theresa, and Stroebel, Johannes. 2020. Social connectedness in urban areas. *Journal of Urban Economics*, **118**, 103264.

Bailey, Michael, Gupta, Abhinav, Hillenbrand, Sebastian, et al. 2021. International trade and social connectedness. *Journal of International Economics*, **129**, 103418.

Baillie, Richard T, Bollerslev, Tim, and Mikkelsen, Hans Ole. 1996. Fractionally integrated generalized autoregressive conditional heteroskedasticity. *Journal of Econometrics*, **74**(1), 3–30.

Bak, Per. 2013. *How nature works: The science of self-organized criticality*. Springer Science & Business Media.

Bak, Per, and Chen, Kan. 1991. Self-organized criticality. *Scientific American*, **264**(1), 46–53.

Bak, Per, and Paczuski, Maya. 1995. Complexity, contingency, and criticality. *Proceedings of the National Acadamy of Sciences of the USA*, **92**, 6689–6696.

Bak, Per, and Sneppen, Kim. 1993. Punctuated equilibrium and criticality in a simple model of evolution. *Physical Review Letters*, **71**(24), 4083–4086.

Bak, Per, and Tang, Chao. 1989. Earthquakes as a self-organized critical phenomenon. *Journal of Geophysical Research: Solid Earth*, **94**(B11), 15635–15637.

Bak, Per, Tang, Chao, and Wiesenfeld, Kurt. 1987. Self-organized criticality: An explanation of the 1/f noise. *Physical Review Letters*, **59**(4), 381–384.

Bak, Per, Tang, Chao, and Wiesenfeld, Kurt. 1988. Self-organized criticality. *Physical Review A*, **38**(1), 364.

Banavar, Jayanth R, Maritan, Amos, and Rinaldo, Andrea. 1999. Size and form in efficient transportation networks. *Nature*, **399**(6732), 130–132.

Banerjee, Abhijit, Chandrasekhar, Arun G, Duflo, Esther, and Jackson, Matthew O. 2013. The diffusion of microfinance. *Science*, **341**(6144), DOI: 10.1126/science.1236498.

Barabási, Albert-László. 2016. *Network science*. Cambridge University Press.

Barabási, Albert-László, and Albert, Réka. 1999. Emergence of scaling in random networks. *Science*, **286**(5439), 509–512.

Battiston, Federico, Nicosia, Vincenzo, and Latora, Vito. 2014. Structural measures for multiplex networks. *Physical Review E*, **89**(3), 032804.

Battiston, Stefano, Puliga, Michelangelo, Kaushik, Rahul, Tasca, Paolo, and Caldarelli, Guido. 2012. Debtrank: Too central to fail? Financial networks, the Fed and systemic risk. *Scientific Reports*, **2**(1), 1–6.

Battiston, Stefano, Farmer, J Doyne, Flache, Andreas, et al. 2016. Complexity theory and financial regulation. *Science*, **351**(6275), 818–819.

Batty, Michael. 2006. Rank clocks. *Nature*, **444**(7119), 592–596.

Baum, Christopher F, and Wiggins, Vince. 2001. Tests for long memory in a time series. *Stata Technical Bulletin*, **10**(57), 362–368.

Bavelas, Alex. 1950. Communication patterns in task-oriented groups. *Journal of the Acoustical Society of America*, **22**(6), 725–730.

Bayarri, M Jésus, and Berger, James O. 2004. The interplay of Bayesian and frequentist analysis. *Statistical Science*, **19**(1), 58–80.

Bayes, Thomas, and Price, Richard. 1763. An essay towards solving a problem in the doctrine of chances. *Philosophical Transactions of the Royal Society of London*, **53**, 370–418.

Benguigui, Lucien, and Blumenfeld-Lieberthal, Efrat. 2007. Beyond the power law: A new approach to analyze city size distributions. *Computers, Environment and Urban Systems*, **31**(6), 648–666.

Benhabib, Jess, Bisin, Alberto, and Zhu, Shenghao. 2011. The distribution of wealth and fiscal policy in economies with finitely lived agents. *Econometrica*, **79**(1), 123–157.

Ben-Hur, Asa, and Weston, Jason. 2010. A user's guide to support vector machines. Pages 223–239 of: *Data mining techniques for the life sciences*. Springer.

Bera, Anil K, and Higgins, Matthew L. 1993. ARCH models: Properties, estimation and testing. *Journal of Economic Surveys*, **7**(4), 305–366.

Berliner, L Mark. 1996. Hierarchical Bayesian time series models. Pages 15–22 of: *Maximum entropy and Bayesian methods*. Springer.

Berry, Donald A, and Hochberg, Yosef. 1999. Bayesian perspectives on multiple comparisons. *Journal of Statistical Planning and Inference*, **82**(1–2), 215–227.

Bianconi, Ginestra. 2018. *Multilayer networks: Structure and function*. Oxford University Press.

Billingsley, Patrick. 2008. *Probability and measure*. John Wiley & Sons.

Birman, Kenneth P. 1993. The process group approach to reliable distributed computing. *Communications of the ACM*, **36**(12), 37–53.

Bishop, Christopher M. 2006. *Pattern recognition and machine learning*. Springer.

Black, Duncan, and Henderson, Vernon. 2003. Urban evolution in the USA. *Journal of Economic Geography*, **3**(4), 343–372.

Bloch, Francis, Jackson, Matthew O, and Tebaldi, Pietro. 2019. Centrality measures in networks. Available at SSRN 2749124.

Boccaletti, Stefano, Bianconi, Ginestra, Criado, Regino, et al. 2014. The structure and dynamics of multilayer networks. *Physics Reports*, **544**(1), 1–122.

Bollerslev, Tim, and Mikkelsen, Hans Ole. 1996. Modeling and pricing long memory in stock market volatility. *Journal of Econometrics*, **73**(1), 151–184.

Bollobas, Bela, and Riordan, Oliver. 2004. The diameter of a scale-free random graph. *Combinatorica*, **24**(1), 5–34.

Bonacich, Phillip. 1972. Factoring and weighting approaches to status scores and clique identification. *Journal of Mathematical Sociology*, **2**(1), 113–120.

Bonacich, Phillip. 1987. Power and centrality: A family of measures. *American Journal of Sociology*, **92**(5), 1170–1182.

Bonacich, Phillip. 1991. Simultaneous group and individual centralities. *Social Networks*, **13**(2), 155–168.

Bose, Indrani, and Chaudhuri, Indranath. 2001. Bacterial evolution and the Bak–Sneppen model. *International Journal of Modern Physics C*, **12**(05), 675–683.

Boser, Bernhard E, Guyon, Isabelle M, and Vapnik, Vladimir N. 1992. A training algorithm for optimal margin classifiers. Pages 144–152 of: *Proceedings of the fifth annual workshop on computational learning theory*.

Box, George EP, Jenkins, Gwilym M, Reinsel, Gregory C, and Ljung, Greta M. 2015. *Time series analysis: Forecasting and control*. Wiley.

Boyd, Stephen, and Vandenberghe, Lieven. 2018. *Introduction to applied linear algebra: Vectors, matrices, and least squares*. Cambridge University Press.

Breiman, Leo. 1996. Bagging predictors. *Machine Learning*, **24**(2), 123–140.

Breza, E, Chandrasekhar, A, Golub, B, and Parvathaneni, A. 2019. Networks in economic development. *Oxford Review of Economic Policy*, **35**(4), 678–721.

Brockwell, Peter J, and Davis, Richard A. 2016. *Introduction to time series and forecasting*. Springer.

Brockwell, Peter J, Davis, Richard A, and Fienberg, Stephen E. 1991. *Time series: Theory and methods*. Springer Science & Business Media.

Bullmore, Ed, and Sporns, Olaf. 2012. The economy of brain network organization. *Nature Reviews Neuroscience*, **13**(5), 336–349.

Caldarelli, Guido. 2007. *Scale-free networks: Complex webs in nature and technology*. Oxford University Press.

Casella, George, and Berger, Roger L. 2021. *Statistical inference*. Cengage Learning.

Chakrabarti, Anindya S. 2012. Effects of the turnover rate on the size distribution of firms: An application of the kinetic exchange models. *Physica A*, **391**(23), 6039–6050.

Chakrabarti, Anindya S. 2017. Scale-free distribution as an economic invariant: A theoretical approach. *Journal of Economic Interaction and Coordination*, **12**(1), 1–26.

Chakrabarti, Anindya S, and Chakrabarti, Bikas K. 2009. Microeconomics of the ideal gas like market models. *Physica A*, **388**(19), 4151–4158.

Chakrabarti, Anindya S, and Ghosh, Diptesh. 2019. Emergence of anti-coordination through reinforcement learning in generalized minority games. *Journal of Economic Interaction and Coordination*, **14**(2), 225–245.

Chakrabarti, Anindya S, and Sinha, Sitabhra. 2016. "Hits" emerge through self-organized coordination in collective response of free agents. *Physical Review E*, **94**(4), 042302.

Chakrabarti, Anindya S, Chakrabarti, Bikas K, Chatterjee, Arnab, and Mitra, Manipushpak. 2009. The Kolkata Paise Restaurant problem and resource utilization. *Physica A*, **388**(12), 2420–2426.

Chakrabarti, Anindya S, Chatterjee, Arnab, Nandi, Tushar, Ghosh, Asim, and Chakraborti, Anirban. 2018. Quantifying invariant features of within-group inequality in consumption across groups. *Journal of Economic Interaction and Coordination*, **13**(3), 469–490.

Chakrabarti, Anindya S, Pichl, Lukáš, and Kaizoji, Taisei. 2019. *Network theory and agent-based modeling in economics and finance*. Springer.

Chakrabarti, Bikas K. 2018. Econophysics as conceived by Meghnad Saha. arXiv preprint arXiv:1808.09279.

Chakrabarti, Bikas K, and Marjit, Sugata. 1995. Self-organisation and complexity in simple model systems: Game of life and economics. *Indian Journal of Physics and Proceedings of the Indian Association for the Cultivation of Science-B*, **69**(6), 681.

Chakrabarti, Bikas K, Chatterjee, Arnab, Ghosh, Asim, Mukherjee, Sudip, and Tamir, Boaz. 2017. *Econophysics of the Kolkata Restaurant problem and related games*. Springer.

Chakraborti, Anirban, and Chakrabarti, Bikas K. 2000. Statistical mechanics of money: How saving propensity affects its distribution. *European Physical Journal B*, **17**(1), 167–170.

Chakraborti, Anirban, Challet, Damien, Chatterjee, Arnab, et al. 2015. Statistical mechanics of competitive resource allocation using agent-based models. *Physics Reports*, **552**, 1–25.

Chakravarty, Satya R. 2009. *Inequality, polarization and poverty*. Springer.

Challet, Damien, Marsili, Matteo, and Zhang, Yi-Cheng. 2004. *Minority games: interacting agents in financial markets*. Oxford University Press.

Champernowne, David G. 1953. A model of income distribution. *Economic Journal*, **63**(250), 318–351.

Chandrasekhar, Arun. 2016. Econometrics of network formation. Pages 303–357 of: *The Oxford Handbook of the Economics of Networks*.

Chatfield, Chris, and Collins, Alexander. 1981. *Introduction to multivariate analysis*. CRC Press.

Chatterjee, Arnab, and Chakrabarti, Bikas K. 2007. Kinetic exchange models for income and wealth distributions. *European Physical Journal B*, **60**(2), 135–149.

Chatterjee, Arnab, Chakrabarti, Bikas K, and Manna, SS. 2004. Pareto law in a kinetic model of market with random saving propensity. *Physica A*, **335**(1–2), 155–163.

Chatterjee, Arnab, Sinha, Sitabhra, and Chakrabarti, Bikas K. 2007. Economic inequality: Is it natural? *Current Science*, **92**, 1383–1389.

Chen, Serena H, and Pollino, Carmel A. 2012. Good practice in Bayesian network modelling. *Environmental Modelling & Software*, **37**, 134–145.

Cho, Jin Seo, Greenwood-Nimmo, Matthew, and Shin, Yongcheol. 2021. Recent developments of the autoregressive distributed lag modelling framework. *Journal of Economic Surveys*, DOI: 10.1111/joes.12450.

Choi, Syngjoo, Galeotti, Andrea, and Goyal, Sanjeev. 2017. Trading in networks: theory and experiments. *Journal of the European Economic Association*, **15**(4), 784–817.

Clauset, Aaron, Young, Maxwell, and Gleditsch, Kristian Skrede. 2007. On the frequency of severe terrorist events. *Journal of Conflict Resolution*, **51**(1), 58–87.

Clauset, Aaron, Shalizi, Cosma Rohilla, and Newman, Mark EJ. 2009. Power-law distributions in empirical data. *SIAM review*, **51**(4), 661–703.

Cohen, Reuven, and Havlin, Shlomo. 2003. Scale-free networks are ultrasmall. *Physical Review Letters*, **90**(5), 058701.

Conrad, Christian, and Haag, Berthold R. 2006. Inequality constraints in the fractionally integrated GARCH model. *Journal of Financial Econometrics*, **4**(3), 413–449.

Cont, Rama. 2001. Empirical properties of asset returns: Stylized facts and statistical issues. *Quantitative Finance*, **1**(2), 223–236.

Cormen, Thomas H. 2009. *Introduction to algorithms*. MIT Press.

Costenbader, Elizabeth, and Valente, Thomas W. 2003. The stability of centrality measures when networks are sampled. *Social Networks*, **25**(4), 283–307.

Cox, David R. 1992. Causality: Some statistical aspects. *Journal of the Royal Statistical Society: Series A (Statistics in Society)*, **155**(2), 291–301.

Cox, David R, and Wermuth, Nanny. 2004. Causality: A statistical view. *International Statistical Review*, **72**(3), 285–305.

Cox, Michael AA, and Cox, Trevor F. 2008. Multidimensional scaling. Pages 315–347 of: *Handbook of data visualization*. Springer.

Cressie, Noel, and Wikle, Christopher K. 2015. *Statistics for spatio-temporal data*. John Wiley & Sons.

Crystal, D. 2002. *Language death*. Cambridge University Press.

Cuniberti, Gianaurelio, Valleriani, Angelo, Vega, José Luis, et al. 2001. Effects of regulation on a self-organized market. *Quantitative Finance*, **1**(3), 332–335.

Das, Kousik, Samanta, Sovan, and Pal, Madhumangal. 2018. Study on centrality measures in social networks: A survey. *Social Network Analysis and Mining*, **8**(1), 1–11.

D'Agostino, Gregorio, and Scala, Antonio. 2014. *Networks of networks: The last frontier of complexity*. Springer.

Davis, Allison, Gardner, Burleigh B, and Gardner, Mary R. 1941. *Deep South: A social anthropological study of caste and class*. University of South Carolina Press.

Day, Richard H. 1994. *Complex economic dynamics*. Volume 1. *An introduction to dynamical systems and market mechanisms*. MIT Press.

De Domenico, Manlio, Solé-Ribalta, Albert, Cozzo, Emanuele, et al. 2013. Mathematical formulation of multilayer networks. *Physical Review X*, **3**(4), 041022.

DeJong, David N, and Dave, Chetan. 2011. *Structural macroeconometrics*. Princeton University Press.

Demeester, Piet, Gryseels, Michael, Autenrieth, Achim, et al. 1999. Resilience in multilayer networks. *IEEE Communications Magazine*, **37**(8), 70–76.

De Paula, Áureo. 2020. Econometric models of network formation. *Annual Review of Economics*, **12**, 775–799.

Deville, Pierre, Linard, Catherine, Martin, Samuel, et al. 2014. Dynamic population mapping using mobile phone data. *Proceedings of the National Academy of Sciences*, **111**(45), 15888–15893.

de Sola Pool, Ithiel, and Kochen, Manfred. 1978. Contacts and influence. *Social Networks*, **1**(1), 5–51.

Dhar, Deepak. 1990. Self-organized critical state of sandpile automaton models. *Physical Review Letters*, **64**(14), 1613–1616.

Dhar, Deepak. 1999. Studying self-organized criticality with exactly solved models. arXiv preprint cond-mat/9909009.

Dhar, Deepak. 2006. Theoretical studies of self-organized criticality. *Physica A*, **369**(1), 29–70.

Di Guilmi, Corrado, Gallegati, Mauro, and Landini, Simone. 2017. *Interactive macroeconomics: Stochastic aggregate dynamics with heterogeneous and interacting agents*. Cambridge University Press.

Di Matteo, Tiziana. 2007. Multi-scaling in finance. *Quantitative Finance*, **7**(1), 21–36.

Di Matteo, Tiziana, Aste, Tomaso, and Dacorogna, Michel M. 2005. Long-term memories of developed and emerging markets: Using the scaling analysis to characterize their stage of development. *Journal of Banking & Finance*, **29**(4), 827–851.

Dickey, David A, and Fuller, Wayne A. 1979. Distribution of the estimators for autoregressive time series with a unit root. *Journal of the American Statistical Association*, **74**(366a), 427–431.

Diebold, Francis X, and Yılmaz, Kamil. 2015. *Financial and macroeconomic connectedness: A network approach to measurement and monitoring*. Oxford University Press.

Diodati, P, Marchesoni, F, and Piazza, S. 1991. Acoustic emission from volcanic rocks: An example of self-organized criticality. *Physical Review Letters*, **67**(17), 2239–2243.

Dobson, Annette J, and Barnett, Adrian G. 2018. *An introduction to generalized linear models*. CRC press.

Dodds, Peter Sheridan, and Rothman, Daniel H. 2000. Geometry of river networks. I. Scaling, fluctuations, and deviations. *Physical Review E*, **63**, 016115.

Donangelo, R, and Fort, H. 2002. Model for mutation in bacterial populations. *Physical Review Letters*, **89**(3), 038101.

Donath, William E, and Hoffman, Alan J. 2003. Lower bounds for the partitioning of graphs. Pages 437–442 of: *Selected papers of Alan J Hoffman: With commentary*. World Scientific.

Dubayah, Ralph O, and Drake, Jason B. 2000. LIDAR remote sensing for forestry. *Journal of Forestry*, **98**(6), 44–46.

Duda, Richard O, and Hart, Peter E. 1973. *Pattern classification and scene analysis*. Wiley.

Duflo, Esther. 2017. The economist as plumber. *American Economic Review*, **107**(5), 1–26.

Duncan, David B. 1955. Multiple range and multiple F tests. *Biometrics*, **11**(1), 1–42.

Dunne, Jennifer A, Williams, Richard J, and Martinez, Neo D. 2002. Food-web structure and network theory: The role of connectance and size. *Proceedings of the National Academy of Sciences*, **99**(20), 12917–12922.

Easley, David, and Kleinberg, Jon. 2010. *Networks, crowds, and markets: Reasoning about a highly connected world*. Cambridge University Press.

Eeckhout, Jan. 2004. Gibrat's law for (all) cities. *American Economic Review*, **94**(5), 1429–1451.

Efron, Bradley. 1979. Computers and the theory of statistics: Thinking the unthinkable. *SIAM Review*, **21**(4), 460–480.

Efron, Bradley. 1981. Nonparametric estimates of standard error: The jackknife, the bootstrap and other methods. *Biometrika*, **68**(3), 589–599.

Efron, Bradley. 1982. *The jackknife, the bootstrap and other resampling plans*. SIAM.

Efron, Bradley. 1986. Why isn't everyone a Bayesian? *American Statistician*, **40**(1), 1–5.

Efron, Bradley. 1998. RA Fisher in the 21st century. *Statistical Science*, 95–114.

Efron, Bradley. 2005. Bayesians, frequentists, and scientists. *Journal of the American Statistical Association*, **100**(469), 1–5.

Efron, Bradley, and Hastie, Trevor. 2016. *Computer age statistical inference*. Cambridge University Press.

Elhabian, Shireen, and Farag, Aly. 2009. *A tutorial on data reduction: Principal component analysis theoretical discussion*. Technical report. Computer Vision and Image Processing Laboratory, University of Louisville.

Elliott, Graham, and Timmermann, Allan. 2016. *Economic forecasting*. Princeton University Press.

Elliott, Graham, Rothenberg, Thomas J, and Stock, James H. 1996. Efficient tests for an autoregressive unit root. *Econometrica*, **64**(4), 813–836.

Elmer, Franz-Josef. 1997. Self-organized criticality with complex scaling exponents in the train model. *Physical Review E*, **56**(6), R6225.

Enders, Walter. 2008. *Applied econometric time series*. John Wiley & Sons.

Engle, Robert F, and Bollerslev, Tim. 1986. Modelling the persistence of conditional variances. *Econometric Reviews*, **5**(1), 1–50.

Erdős, Paul, and Rényi, Alfréd. 1959. On random graphs, I. *Publicationes Mathematicae (Debrecen)*, **6**, 290–297.

Ester, Martin, Kriegel, Hans-Peter, Sander, Jörg, et al. 1996. A density-based algorithm for discovering clusters in large spatial databases with noise. Pages 226–231 of: *KDD*, vol. 96.

Euler, L. 1736. Solutio problematis ad geometriam situs pertinensis. *Commentarii academiae scientiarum Petropolitanae*, **8**(1741), 128–140.

Everitt, Brian, and Hothorn, Torsten. 2011. *An introduction to applied multivariate analysis with R*. Springer Science & Business Media.

Fama, Eugene F, and French, Kenneth R. 1993. Common risk factors in the returns on stocks and bonds. *Journal of Financial Economics*, **33**(1), 3–56.

Farine, Damien R, and Whitehead, Hal. 2015. Constructing, conducting and interpreting animal social network analysis. *Journal of Animal Ecology*, **84**(5), 1144–1163.

Fazio, Giorgio, and Modica, Marco. 2015. Pareto or log-normal? Best fit and truncation in the distribution of all cities. *Journal of Regional Science*, **55**(5), 736–756.

Feller, William. 2008a. *An introduction to probability theory and its applications*. Vol. 1. John Wiley & Sons.

Feller, William. 2008b. *An introduction to probability theory and its applications*. Vol 2. John Wiley & Sons.

Ferguson, Niall. 2017. *The square and the tower: Networks, hierarchies and the struggle for global power*. Penguin.

Fiedler, Miroslav. 1973. Algebraic connectivity of graphs. *Czechoslovak Mathematical Journal*, **23**(2), 298–305.

Fisher, Ronald A. 1936. The use of multiple measurements in taxonomic problems. *Annals of Eugenics*, **7**(2), 179–188.

Fornito, Alex, Zalesky, Andrew, and Bullmore, Edward. 2016. *Fundamentals of brain network analysis*. Academic Press.

Fortunato, Santo, and Hric, Darko. 2016. Community detection in networks: A user guide. *Physics Reports*, **659**, 1–44.

Freeman, Linton C. 1977. A set of measures of centrality based on betweenness. *Sociometry*, **40**(1), 35–41.

Freeman, Linton C. 1978. Centrality in social networks conceptual clarification. *Social Networks*, **1**(3), 215–239.

Freund, Yoav, and Schapire, Robert E. 1996. Experiments with a new boosting algorithm. Pages 148–156 of: *Machine learning: Proceedings of the thirteenth international conference*, vol. 96. Citeseer.

Fung, Glenn M, and Mangasarian, Olvi L. 2005. Multicategory proximal support vector machine classifiers. *Machine Learning*, **59**(1–2), 77–97.

Gabaix, Xavier. 1999. Zipf's law for cities: An explanation. *Quarterly Journal of Economics*, **114**(3), 739–767.

Gabaix, Xavier. 2014. A sparsity-based model of bounded rationality. *Quarterly Journal of Economics*, **129**(4), 1661–1710.

Gabaix, Xavier. 2016. Power laws in economics: An introduction. *Journal of Economic Perspectives*, **30**(1), 185–206.

Gabaix, Xavier, and Ibragimov, Rustam. 2011. Rank – 1/2: A simple way to improve the OLS estimation of tail exponents. *Journal of Business & Economic Statistics*, **29**(1), 24–39.

Gabaix, Xavier, and Ioannides, Yannis M. 2004. The evolution of city size distributions. Pages 2341–2378 of: *Handbook of regional and urban economics*, vol. 4. Elsevier.

Galam, Serge, and Mauger, Alain. 2003. On reducing terrorism power: A hint from physics. *Physica A*, **323**, 695–704.

Gan, Li, Li, Dong, and Song, Shunfeng. 2006. Is the Zipf law spurious in explaining city-size distributions? *Economics Letters*, **92**(2), 256–262.

Gangopadhyay, Kausik, and Basu, B. 2009. City size distributions for India and China. *Physica A*, **388**(13), 2682–2688.

Gangopadhyay, Kausik, and Basu, Banasri. 2013. Evolution of Zipf's law for Indian urban agglomerations vis-à-vis Chinese urban agglomerations. Pages 119–129 of: *Econophysics of systemic risk and network dynamics*. Springer.

Gao, Jianbo, Hu, Jing, Tung, Wen-Wen, et al. 2006. Assessment of long-range correlation in time series: How to avoid pitfalls. *Physical Review E*, **73**(1), 016117.

Garlaschelli, Diego, and Loffredo, Maria I. 2004. Patterns of link reciprocity in directed networks. *Physical Review Letters*, **93**(26), 268701.

Garrido, Pablo, Lovejoy, Shaun, and Schertzer, Daniel. 1996. Multifractal processes and self-organized criticality in the large-scale structure of the universe. *Physica A*, **225**(3–4), 294–311.

Gauss, Carl Friedrich. 1809. *Theoria motus corporum coelestium in sectionibus conicis solem ambientium*. Vol. 7. Perthes et Besser.

Gelfand, Alan E. 2012. Hierarchical modeling for spatial data problems. *Spatial Statistics*, **1**, 30–39.

Gelfand, Alan E, and Banerjee, Sudipto. 2017. Bayesian modeling and analysis of geostatistical data. *Annual Review of Statistics and its Application*, **4**, 245–266.

Gell-Mann, Murray. 2002. What is complexity? Pages 13–24 of: *Complexity and industrial clusters*. Springer.

Gelman, A, Carlin, J, Stern, HS, and Rubin, DB. 2004. *Bayesian data analysis*. CRC Press.

Gelman, Andrew, and Shalizi, Cosma Rohilla. 2013. Philosophy and the practice of Bayesian statistics. *British Journal of Mathematical and Statistical Psychology*, **66**(1), 8–38.

Gelman, Andrew, Carlin, John B, Stern, Hal S, Dunson, David B, Vehtari, Aki, and Rubin, Donald B. 2013. *Bayesian data analysis*. CRC press.

George, Edward I, and McCulloch, Robert E. 1993. Variable selection via Gibbs sampling. *Journal of the American Statistical Association*, **88**(423), 881–889.

Ghemawat, Sanjay, Gobioff, Howard, and Leung, Shun-Tak. 2003. The Google file system. Pages 29–43 of: *Proceedings of the nineteenth ACM symposium on operating systems principles*.

Ghosh, Asim, Chatterjee, Arnab, Inoue, Jun-ichi, and Chakrabarti, Bikas K. 2016. Inequality measures in kinetic exchange models of wealth distributions. *Physica A*, **451**, 465–474.

Ghosh, Diptesh, and Chakrabarti, Anindya S. 2017. Emergence of distributed coordination in the Kolkata paise restaurant problem with finite information. *Physica A*, **483**, 16–24.

Ghysels, Eric, Sinko, Arthur, and Valkanov, Rossen. 2007. MIDAS regressions: Further results and new directions. *Econometric Reviews*, **26**(1), 53–90.

Gibrat, R. 1931. *Les inégalités économiques, Librairie du Recueil Sirey, Paris*. Vol. 169. Librairie du Recueil Sirey.

Gilbert, Edgar N. 1959. Random graphs. *Annals of Mathematical Statistics*, **30**(4), 1141–1144.

Gil-Mendieta, Jorge, and Schmidt, Samuel. 1996. The political network in Mexico. *Social Networks*, **18**(4), 355–381.

Gintis, Herbert. 2009. *Game theory evolving*. Princeton University Press.

Girvan, Michelle, and Newman, Mark EJ. 2002. Community structure in social and biological networks. *Proceedings of the National Academy of Sciences*, **99**(12), 7821–7826.

Gittins, John C. 1979. Bandit processes and dynamic allocation indices. *Journal of the Royal Statistical Society: Series B (Methodological)*, **41**(2), 148–164.

Gomez, Sergio, Diaz-Guilera, Albert, Gomez-Gardenes, Jesus, et al. 2013. Diffusion dynamics on multiplex networks. *Physical Review Letters*, **110**(2), 028701.

Gómez-Déniz, Emilio, and Calderín-Ojeda, Enrique. 2015. On the use of the Pareto ArcTan distribution for describing city size in Australia and New Zealand. *Physica A*, **436**, 821–832.

González-Val, Rafael. 2010. The evolution of US city size distribution from a long-term perspective (1900–2000). *Journal of Regional Science*, **50**(5), 952–972.

González-Val, Rafael. 2012. A nonparametric estimation of the local Zipf exponent for all US cities. *Environment and Planning B: Planning and Design*, **39**(6), 1119–1130.

González-Val, Rafael, Lanaspa, Luis, and Sanz-Gracia, Fernando. 2014. New evidence on Gibrat's law for cities. *Urban Studies*, **51**(1), 93–115.

Gould, Stephen Jay, and Eldredge, Niles. 1977. Punctuated equilibria: The tempo and mode of evolution reconsidered. *Paleobiology*, **3**(2), 115–151.

Gower, John C. 1966. Some distance properties of latent root and vector methods used in multivariate analysis. *Biometrika*, **53**(3–4), 325–338.

Goyal, Sanjeev. 2012. *Connections: An introduction to the economics of networks*. Princeton University Press.

Granero, MA Sánchez, Segovia, JE Trinidad, and Pérez, J García. 2008. Some comments on Hurst exponent and the long memory processes on capital markets. *Physica A: Statistical Mechanics and its Applications*, **387**(22), 5543–5551.

Granger, Clive WJ, and Joyeux, Roselyne. 1980. An introduction to long-memory time series models and fractional differencing. *Journal of Time Series Analysis*, **1**(1), 15–29.

Granovetter, Mark S. 1973. The strength of weak ties. *American Journal of Sociology*, **78**(6), 1360–1380.

Graves, Timothy, Gramacy, Robert, Watkins, Nicholas, and Franzke, Christian. 2017. A brief history of long memory: Hurst, Mandelbrot and the road to ARFIMA, 1951–1980. *Entropy*, **19**(9), 437, DOI: 10.3390/e19090437.

Greene, William H. 2003. *Econometric analysis*. Pearson Education India.

Guha, Pritha, Bansal, Avijit, Guha, Apratim, and Chakrabarti, Anindya S. 2021. Gravity and depth of social media networks. *Journal of Complex Networks*, **9**(2), cnab016.

Gutierrez-Osuna, Ricardo. 2020. *Linear discriminants analysis*. Technical report. Texas A&M University.

Haldane, Andrew G, and May, Robert M. 2011. Systemic risk in banking ecosystems. *Nature*, **469**(7330), 351–355.

Hamilton, James D. 1994. *Time series analysis*. Princeton. University Press.

Hansen, Lars Kai, and Salamon, Peter. 1990. Neural network ensembles. *IEEE Transactions on Pattern Analysis and Machine Intelligence*, **12**(10), 993–1001.

Hansen, Lars Peter. 1982. Large sample properties of generalized method of moments estimators. *Econometrica: Journal of the Econometric Society*, 1029–1054.

Harary, Frank. 1953. On the notion of balance of a signed graph. *Michigan Mathematical Journal*, **2**(2), 143–146.

Hastie, Trevor, Tibshirani, Robert, and Friedman, Jerome H. 2009. *The elements of statistical learning: Data mining, inference, and prediction*. Springer.

Hastie, Trevor J, and Tibshirani, Robert J. 1990. *Generalized additive models*. CRC press.

Haykin, Simon. 1994. *Neural networks: A comprehensive foundation*. Prentice Hall PTR.

Heinsalu, Els, Patriarca, Marco, and Léonard, Jean Léo. 2014. The role of bilinguals in language competition. *Advances in Complex Systems*, **17**(01), 1450003.

Herr, Bruce W, Ke, Weimao, Hardy, Elisha, and Borner, Katy. 2007. Movies and actors: Mapping the internet movie database. Pages 465–469 of: *2007 11th International Conference Information Visualization (IV'07)*. IEEE.

Hidalgo, César A, and Hausmann, Ricardo. 2009. The building blocks of economic complexity. *Proceedings of the National Academy of Sciences*, **106**(26), 10570–10575.

Hoerl, Arthur E, and Kennard, Robert W. 1970. Ridge regression: Biased estimation for nonorthogonal problems. *Technometrics*, **12**(1), 55–67.

Hogg, Robert V, McKean, Joseph, and Craig, Allen T. 2005. *Introduction to mathematical statistics*. Pearson Education.

Holland, Paul W, and Leinhardt, Samuel. 1974. *The statistical analysis of local structure in social networks*. Technical report. National Bureau of Economic Research.

Holland, Paul W, and Leinhardt, Samuel. 1975. The statistical analysis of local structure in social networks. *Sociological Methodology*, David Heise, ed. Jossey-Bass.

Holland, Paul W, and Leinhardt, Samuel. 1976. Local structure in social networks. *Sociological Methodology*, **7**, 1–45.

Holland, Paul W, and Leinhardt, Samuel. 1977. A method for detecting structure in sociometric data. Pages 411–432 of: *Social Networks*. Elsevier.

Holme, Petter, and Saramäki, Jari. 2012. Temporal networks. *Physics Reports*, **519**(3), 97–125.

Holme, Petter, and Saramäki, Jari. 2019. *Temporal Network Theory*. Springer.

Hosking, Jonathan RM. 1984. Modeling persistence in hydrological time series using fractional differencing. *Water Resources Research*, **20**(12), 1898–1908.

Hotelling, Harold. 1933. Analysis of a complex of statistical variables into principal components. *Journal of Educational Psychology*, **24**(6), 417–441.

Hu, Yu-Pin, and Tsay, Ruey S. 2014. Principal volatility component analysis. *Journal of Business & Economic Statistics*, **32**(2), 153–164.

Huang, Yue, and Gao, Xuedong. 2014. Clustering on heterogeneous networks. *Wiley Interdisciplinary Reviews: Data Mining and Knowledge Discovery*, **4**(3), 213–233.

Huberman, Bernardo A. 2001. *The laws of the Web*. MIT Press.

Hummon, Norman P, and Dereian, Patrick. 1989. Connectivity in a citation network: The development of DNA theory. *Social Networks*, **11**(1), 39–63.

Imbens, Guido W, and Rubin, Donald B. 2015. *Causal inference in statistics, social, and biomedical sciences*. Cambridge University Press.

Ioannides, Yannis, and Skouras, Spyros. 2013. US city size distribution: Robustly Pareto, but only in the tail. *Journal of Urban Economics*, **73**(1), 18–29.

Ishwaran, Hemant, and Rao, J Sunil. 2005. Spike and slab variable selection: Frequentist and Bayesian strategies. *Annals of Statistics*, **33**(2), 730–773.

Ivashkevich, EV, and Priezzhev, Vyatcheslav B. 1998. Introduction to the sandpile model. *Physica A*, **254**(1–2), 97–116.

Jackson, Matthew O. 2010. *Social and economic networks*. Princeton University Press.

Jackson, Matthew O. 2019. The role of theory in an age of design and big data. Pages 523–530 of: *The Future of Economic Design*. Springer.

James, Gareth, Witten, Daniela, Hastie, Trevor, and Tibshirani, Robert. 2013. *An introduction to statistical learning*. Springer.

Jeong, Hawoong, Tombor, Bálint, Albert, Réka, Oltvai, Zoltan N, and Barabási, Albert-László. 2000. The large-scale organization of metabolic networks. *Nature*, **407**(6804), 651–654.

Jeong, Hawoong, Mason, Sean P, Barabási, Albert-László, and Oltvai, Zoltan N. 2001. Lethality and centrality in protein networks. *Nature*, **411**(6833), 41–42.

John, George H, and Langley, Pat. 1995. Estimating continuous distributions in Bayesian classifiers. Pages 338–345 of: *Proceedings of the Eleventh Conference on Uncertainty in Artificial Intelligence*.

Johnson, Richard Arnold, and Wichern, Dean W. 2014. *Applied multivariate statistical analysis*. Pearson.

Johnson, Stephen C. 1967. Hierarchical clustering schemes. *Psychometrika*, **32**(3), 241–254.

Jolliffe, Ian T, and Cadima, Jorge. 2016. Principal component analysis: A review and recent developments. *Philosophical Transactions of the Royal Society A: Mathematical, Physical and Engineering Sciences*, **374**(2065), 20150202.

Jordan, Michael I, and Mitchell, Tom M. 2015. Machine learning: Trends, perspectives, and prospects. *Science*, **349**(6245), 255–260.

Kandler, Anne, and Unger, Roman. 2018. Modeling language shift. Pages 351–373 of: *Diffusive spreading in nature, technology and society*. Springer.

Kanungo, Tapas, Mount, David M, Netanyahu, Nathan S, et al. 2002. An efficient k-means clustering algorithm: Analysis and implementation. *IEEE Transactions on Pattern Analysis and Machine Intelligence*, **24**(7), 881–892.

Karabarbounis, Loukas, and Neiman, Brent. 2014. The global decline of the labor share. *Quarterly Journal of Economics*, **129**(1), 61–103.

Karinthy, Frigyes. 1929. Chain-links. *Everything is different*.

Kauffman, Stuart A. 2000. *Investigations*. Oxford University Press.

Kilian, Lutz, and Lütkepohl, Helmut. 2017. *Structural vector autoregressive analysis*. Cambridge University Press.

Kirman, Alan. 1993. Ants, rationality, and recruitment. *Quarterly Journal of Economics*, **108**(1), 137–156.

Kivelä, Mikko, Arenas, Alex, Barthelemy, Marc, et al. 2014. Multilayer networks. *Journal of Complex Networks*, **2**(3), 203–271.

Koppl, Roger, Devereaux, Abigail, Valverde, Sergi, et al. 2021. Explaining Technology. Available at SSRN 3856338.

Koski, Timo, and Noble, John. 2011. *Bayesian networks: An introduction*. John Wiley & Sons.

Krus, David J, and Fuller, Ellen A. 1982. Computer assisted multicrossvalidation in regression analysis. *Educational and Psychological Measurement*, **42**(1), 187–193.

Kruskal, Joseph B. 1956. On the shortest spanning subtree of a graph and the traveling salesman problem. *Proceedings of the American Mathematical Society*, 7(1), 48–50.

Kruskal, Joseph B. 1964a. Multidimensional scaling by optimizing goodness of fit to a nonmetric hypothesis. *Psychometrika*, **29**(1), 1–27.

Kruskal, Joseph B. 1964b. Nonmetric multidimensional scaling: A numerical method. *Psychometrika*, **29**(2), 115–129.

Kruskal, Joseph B. 1978. *Multidimensional scaling*. Sage.

Kuchler, T, Russel, D, and Stroebel, J. 2021. The geographic spread of COVID-19 correlates with the structure of social networks as measured by Facebook. *Journal of Urban Economics*, 103314.

Kumar, Ashish, Chakrabarti, Anindya S, Chakraborti, Anirban, and Nandi, Tushar. 2021. Distress propagation on production networks: Coarse-graining and modularity of linkages. *Physica A: Statistical Mechanics and its Applications*, **568**, 125714.

Kuncheva, Ludmila I. 2014. *Combining pattern classifiers: Methods and algorithms*. John Wiley & Sons.

Kurtz, Albert K. 1948. A research test of the Rorschach test. *Personnel Psychology*, **1**, 41–51.

Kuyyamudi, Chandrashekar, Chakrabarti, Anindya S, and Sinha, Sitabhra. 2019. Emergence of frustration signals systemic risk. *Physical Review E*, **99**(5), 052306.

Kvam, Paul H, and Vidakovic, Brani. 2007. *Nonparametric statistics with applications to science and engineering.* John Wiley & Sons.

Kwiatkowski, Denis, Phillips, Peter CB, Schmidt, Peter, and Shin, Yongcheol. 1992. Testing the null hypothesis of stationarity against the alternative of a unit root: How sure are we that economic time series have a unit root? *Journal of Econometrics,* **54**(1–3), 159–178.

Ladyman, James, and Wiesner, Karoline. 2020. *What is a complex system?* Yale University Press.

Ladyman, James, Lambert, James, and Wiesner, Karoline. 2013. What is a complex system? *European Journal for Philosophy of Science,* **3**(1), 33–67.

Lafferty, John, and Wasserman, Larry. 2006. Challenges in statistical machine learning. *Statistica Sinica,* **16**(2), 307–322.

Lamport, Leslie. 2019. Time, clocks, and the ordering of events in a distributed system. Pages 179–196 of: *Concurrency: The Works of Leslie Lamport.* ACM Books.

LeBaron, Blake. 2016. Financial price dynamics and agent-based models as inspired by Benoit Mandelbrot. *European Physical Journal – Special Topics,* **225**(17), 3243–3254.

Lee, Jae Kook, Choi, Jihyang, Kim, Cheonsoo, and Kim, Yonghwan. 2014. Social media, network heterogeneity, and opinion polarization. *Journal of Communication,* **64**(4), 702–722.

Legendre, Adrien Marie. 1805. *Nouvelles méthodes pour la détermination des orbites des comètes.* F. Didot.

Lewis, Kevin, Kaufman, Jason, Gonzalez, Marco, Wimmer, Andreas, and Christakis, Nicholas. 2008. Tastes, ties, and time: A new social network dataset using Facebook.com. *Social Networks,* **30**(4), 330–342.

Li, Aming, Cornelius, Sean P, Liu, Y-Y, Wang, Long, and Barabási, A-L. 2017. The fundamental advantages of temporal networks. *Science,* **358**(6366), 1042–1046.

Li, Cong, Li, Qian, Van Mieghem, Piet, Stanley, H Eugene, and Wang, Huijuan. 2015. Correlation between centrality metrics and their application to the opinion model. *European Physical Journal B,* **88**(3), 1–13.

Li, Shujuan, and Sui, Daniel. 2013. Pareto's law and sample size: A case study of China's urban system 1984–2008. *GeoJournal,* **78**(4), 615–626.

Likas, Aristidis, Vlassis, Nikos, and Verbeek, Jakob J. 2003. The global k-means clustering algorithm. *Pattern Recognition,* **36**(2), 451–461.

Liljeros, Fredrik, Edling, Christofer R, Stanley, H Eugene, Åberg, Y, and Amaral, Luis AN. 2003. Sexual contacts and epidemic thresholds. *Nature,* **423**(6940), 606.

Lindley, DV, and Scott, WF, 1995. *New Cambridge Statistical Tables.* 2nd edn. Cambridge University Press.

Luce, R Duncan, and Perry, Albert D. 1949. A method of matrix analysis of group structure. *Psychometrika,* **14**(2), 95–116.

Luckstead, Jeff, and Devadoss, Stephen. 2014. A comparison of city size distributions for China and India from 1950 to 2010. *Economics Letters,* **124**(2), 290–295.

Luckstead, Jeff, and Devadoss, Stephen. 2017. Pareto tails and lognormal body of US cities size distribution. *Physica A,* **465**, 573–578.

Lusseau, David. 2003. The emergent properties of a dolphin social network. *Proceedings of the Royal Society of London B: Biological Sciences,* **270**(Suppl 2), S186–S188.

Lux, Thomas. 2018. Estimation of agent-based models using sequential Monte Carlo methods. *Journal of Economic Dynamics and Control,* **91**, 391–408.

Lux, Thomas, and Marchesi, Michele. 1999. Scaling and criticality in a stochastic multi-agent model of a financial market. *Nature,* **397**(6719), 498–500.

MacQueen, James. 1967. Some methods for classification and analysis of multivariate observations. Pages 281–297 of: *Proceedings of the fifth Berkeley symposium on mathematical statistics and probability*, vol. 1.

Malacarne, Luis Carlos, Mendes, RS, and Lenzi, Ervin Kaminski. 2001. q-Exponential distribution in urban agglomeration. *Physical Review E*, **65**(1), 017106.

Malamud, Bruce D, Morein, Gleb, and Turcotte, Donald L. 1998. Forest fires: An example of self-organized critical behavior. *Science*, **281**(5384), 1840–1842.

Malo, Pekka, Sinha, Ankur, Korhonen, Pekka, Wallenius, Jyrki, and Takala, Pyry. 2014. Good debt or bad debt: Detecting semantic orientations in economic texts. *Journal of the Association for Information Science and Technology*, **65**(4), 782–796.

Mandelbrot, Benoit. 1960. The Pareto-Levy law and the distribution of income. *International Economic Review*, **1**(2), 79–106.

Mandelbrot, Benoit, and Hudson, Richard L. 2007. *The misbehavior of markets: A fractal view of financial turbulence*. Basic Books.

Mandelbrot, Benoit B. 1997. The variation of certain speculative prices. Pages 371–418 of: *Fractals and scaling in finance*. Springer.

Mandelbrot, Benoit B. 2013. *Fractals and scaling in finance: Discontinuity, concentration, risk. Selecta volume E*. Springer Science & Business Media.

Mandelbrot, Benoit B, and Wallis, James R. 1969. Robustness of the rescaled range R/S in the measurement of noncyclic long run statistical dependence. *Water Resources Research*, **5**(5), 967–988.

Mantegna, Rosario N. 1999. Hierarchical structure in financial markets. *European Physical Journal B – Condensed Matter and Complex Systems*, **11**(1), 193–197.

Mantegna, Rosario Nunzio, and Stanley, H. Eugene. 2007. *An introduction to econophysics: Correlations and complexity in finance*. Cambridge University Press.

Maritan, Amos, Rinaldo, Andrea, Rigon, Riccardo, Giacometti, Achille, and Rodríguez-Iturbe, Ignacio. 1996. Scaling laws for river networks. *Physical Review E*, **53**(2), 1510–1515.

Marsili, Matteo, and Zhang, Yi-Cheng. 1998. Interacting individuals leading to Zipf's law. *Physical Review Letters*, **80**(12), 2741–2744.

Massara, Guido Previde, Di Matteo, Tiziana, and Aste, Tomaso. 2017. Network filtering for big data: Triangulated maximally filtered graph. *Journal of Complex Networks*, **5**(2), 161–178.

May, Robert M. 1972. Will a large complex system be stable? *Nature*, **238**(5364), 413–414.

May, Robert M, Levin, Simon A, and Sugihara, George. 2008. Ecology for bankers. *Nature*, **451**(7181), 893–894.

McCullagh, Peter. 2018. *Generalized linear models*. Routledge.

McGill, Robert, Tukey, John W, and Larsen, Wayne A. 1978. Variations of box plots. *American Statistician*, **32**(1), 12–16.

Menczer, Filippo, Fortunato, Santo, and Davis, Clayton A. 2020. *A first course in network science*. Cambridge University Press.

Menichetti, Giulia, Remondini, Daniel, Panzarasa, Pietro, Mondragón, Raúl J, and Bianconi, Ginestra. 2014. Weighted multiplex networks. *PLOS One*, **9**(6), e97857.

Milo, Ron, Shen-Orr, Shai, Itzkovitz, Shalev et al. 2002. Network motifs: Simple building blocks of complex networks. *Science*, **298**(5594), 824–827.

Minett, James W, and Wang, William SY. 2008. Modelling endangered languages: The effects of bilingualism and social structure. *Lingua*, **118**(1), 19–45.

Mishra, Abinash, Srivastava, Pranjal, and Chakrabarti, Anindya S. 2021. "Too central to fail" firms in bi-layered financial networks: Linkages in the US corporate bond and stock markets. *Quantitative Finance*, **22**(5), 943–971.

Mitchell, Tom M. 1997. *Machine learning*. McGraw-hill New York.

Mitzenmacher, Michael. 2004. A brief history of generative models for power law and lognormal distributions. *Internet Mathematics*, **1**(2), 226–251.

Mohri, Mehryar, Rostamizadeh, Afshin, and Talwalkar, Ameet. 2018. *Foundations of machine learning*. MIT Press.

Moon, John W, and Moser, Leo. 1965. On cliques in graphs. *Israel Journal of Mathematics*, **3**(1), 23–28.

Moreira, Andre A, Andrade Jr, José S, Herrmann, Hans J, and Indekeu, Joseph O. 2009. How to make a fragile network robust and vice versa. *Physical Review Letters*, **102**(1), 018701.

Moreno, Jacob Levy. 1953. *Who shall survive? Foundations of sociometry, group psychotherapy and sociodrama*. Beacon House.

Mosier, Charles I. 1951. I. Problems and designs of cross-validation 1. *Educational and Psychological Measurement*, **11**(1), 5–11.

Motter, Adilson E, Myers, Seth A, Anghel, Marian, and Nishikawa, Takashi. 2013. Spontaneous synchrony in power-grid networks. *Nature Physics*, **9**(3), 191–197.

Mucha, Peter J, Richardson, Thomas, Macon, Kevin, Porter, Mason A, and Onnela, Jukka-Pekka. 2010. Community structure in time-dependent, multiscale, and multiplex networks. *Science*, **328**(5980), 876–878.

Mullainathan, Sendhil, and Spiess, Jann. 2017. Machine learning: An applied econometric approach. *Journal of Economic Perspectives*, **31**(2), 87–106.

Munshi, Kaivan. 2003. Networks in the modern economy: Mexican migrants in the US labor market. *The Quarterly Journal of Economics*, **118**(2), 549–599.

Munshi, Kaivan, and Rosenzweig, Mark. 2016. Networks and misallocation: Insurance, migration, and the rural–urban wage gap. *American Economic Review*, **106**(1), 46–98.

Murphy, Kevin P. 2012. *Machine learning: A probabilistic perspective*. MIT Press.

Nagurney, Anna, Pan, Jie, and Zhao, Lan. 1992. Human migration networks. *European Journal of Operational Research*, **59**(2), 262–274.

Namatame, Akira, and Chen, Shu-Heng. 2016. *Agent-based modeling and network dynamics*. Oxford University Press.

Nelder, John Ashworth, and Wedderburn, Robert WM. 1972. Generalized linear models. *Journal of the Royal Statistical Society: Series A (General)*, **135**(3), 370–384.

Nettle, Daniel. 1998. Explaining global patterns of language diversity. *Journal of Anthropological Archaeology*, **17**(4), 354–374.

Neusser, K. 2016. *Time series econometrics*. Springer.

Newman, Mark EJ. 2000. Models of the small world. *Journal of Statistical Physics*, **101**(3–4), 819–841.

Newman, Mark EJ. 2001. Scientific collaboration networks. I. Network construction and fundamental results. *Physical Review E*, **64**(1), 016131.

Newman, Mark EJ. 2003. The structure and function of complex networks. *SIAM Review*, **45**(2), 167–256.

Newman, Mark EJ. 2004a. Coauthorship networks and patterns of scientific collaboration. *Proceedings of the National Academy of Sciences*, **101**(suppl 1), 5200–5205.

Newman, Mark EJ. 2004b. Detecting community structure in networks. *European Physical Journal B*, **38**(2), 321–330.

Newman, Mark EJ. 2005. Power laws, Pareto distributions and Zipf's law. *Contemporary Physics*, **46**(5), 323–351.

Newman, Mark EJ. 2010. *Networks: An introduction*. Oxford University Press.

Newman, Mark EJ, and Girvan, Michelle. 2004. Finding and evaluating community structure in networks. *Physical Review E*, **69**(2), 026113.

Newman, Mark EJ, and Park, Juyong. 2003. Why social networks are different from other types of networks. *Physical Review E*, **68**(3), 036122.

Newman, Mark EJ, Strogatz, Steven H, and Watts, Duncan J. 2001. Random graphs with arbitrary degree distributions and their applications. *Physical Review E*, **64**(2), 026118.

Newman, Mark EJ, Forrest, Stephanie, and Balthrop, Justin. 2002a. Email networks and the spread of computer viruses. *Physical Review E*, **66**(3), 035101.

Newman, Mark EJ, Watts, Duncan J, and Strogatz, Steven H. 2002b. Random graph models of social networks. *Proceedings of the National Academy of Sciences*, **99**(suppl 1), 2566–2572.

Newman, MEJ. 1996. Self-organized criticality, evolution and the fossil extinction record. *Proceedings of the Royal Society of London. Series B: Biological Sciences*, **263**(1376), 1605–1610.

Newman, MEJ. 1997. Evidence for self-organized criticality in evolution. *Physica D*, **107**(2–4), 293–296.

Ng, Serena, and Perron, Pierre. 2001. Lag length selection and the construction of unit root tests with good size and power. *Econometrica*, **69**(6), 1519–1554.

Nicosia, Vincenzo, Bianconi, Ginestra, Latora, Vito, and Barthelemy, Marc. 2013. Growing multiplex networks. *Physical Review Letters*, **111**(5), 058701.

Noble, William S. 2006. What is a support vector machine? *Nature Biotechnology*, **24**(12), 1565–1567.

Nocedal, Jorge, and Wright, Stephen. 2006. *Numerical optimization*. Springer Science & Business Media.

Nowé, Ann, Vrancx, Peter, and Hauwere, Yann-Michaël De. 2012. Game theory and multi-agent reinforcement learning. Pages 441–470 of: *Reinforcement learning*. Springer.

O'Hara, Robert B, and Sillanpää, Mikko J. 2009. A review of Bayesian variable selection methods: What, how and which. *Bayesian Analysis*, **4**(1), 85–117.

Onnela, J-P, Saramäki, Jari, Hyvönen, Jorkki, et al. 2007. Structure and tie strengths in mobile communication networks. *Proceedings of the National Academy of Sciences*, **104**(18), 7332–7336.

Opitz, David, and Maclin, Richard. 1999. Popular ensemble methods: An empirical study. *Journal of Artificial Intelligence Research*, **11**, 169–198.

Osborne, Martin J. 2004. *An introduction to game theory*. Oxford University Press.

Page, Scott. 2010. *Diversity and complexity*. Princeton University Press.

Pan, Raj Kumar, Chatterjee, Nivedita, and Sinha, Sitabhra. 2010. Mesoscopic organization reveals the constraints governing Caenorhabditis elegans nervous system. *PLOS One*, **5**(2), e9240.

Pareto, V. 1897. *Cours D'Economie Politique*. Vol. 2. University of Lausanne.

Pearl, Judea. 2009. *Causality*. Cambridge University Press.

Pearson, Karl. 1901. LIII. On lines and planes of closest fit to systems of points in space. *London, Edinburgh, and Dublin Philosophical Magazine and Journal of Science*, **2**(11), 559–572.

Peng, C-K, Buldyrev, Sergej V, Goldberger, Ary L, et al. 1992. Long-range correlations in nucleotide sequences. *Nature*, **356**(6365), 168–170.

Peng, C-K, Buldyrev, Sergey V, Havlin, Shlomo, et al. 1994. Mosaic organization of DNA nucleotides. *Physical Review E*, **49**(2), 1685–1689.

Peng, C-K, Havlin, S, Hausdorff, JM, et al. 1995. Fractal mechanisms and heart rate dynamics: Long-range correlations and their breakdown with disease. *Journal of Electrocardiology*, **28**, 59–65.

Pham, Duc Truong, Dimov, Stefan S, and Nguyen, Chi D. 2005. Selection of K in K-means clustering. *Proceedings of the Institution of Mechanical Engineers, Part C: Journal of Mechanical Engineering Science*, **219**(1), 103–119.

Phillips, Peter CB, and Perron, Pierre. 1988. Testing for a unit root in time series regression. *Biometrika*, **75**(2), 335–346.

Picoli, S, del Castillo-Mussot, M, Ribeiro, HV, Lenzi, EK, and Mendes, RS. 2014. Universal bursty behaviour in human violent conflicts. *Scientific Reports*, **4**(1), 1–3.

Piketty, Thomas. 2014. *Capital in the twenty-first century*. Harvard University Press.

Polikar, Robi. 2006. Ensemble based systems in decision making. *IEEE Circuits and Systems Magazine*, **6**(3), 21–45.

Porter, Mason A. 2018. What is ... a Multilayer Network? *Notices of the AMS*, **65**(11), 1419–1423.

Price, Derek de Solla. 1976. A general theory of bibliometric and other cumulative advantage processes. *Journal of the American Society for Information Science*, **27**(5), 292–306.

Price, Derek J. De Solla. 1965. Networks of scientific papers. *Science*, **149**(3683), 510–515.

Prim, Robert Clay. 1957. Shortest connection networks and some generalizations. *Bell Labs Technical Journal*, **36**(6), 1389–1401.

Quevedo, David Santiago, and Quimbay, Carlos José. 2020. Non-conservative kinetic model of wealth exchange with saving of production. *European Physical Journal B*, **93**(10), 1–12.

Quinlan, J Ross. 2014. *C4.5: Programs for machine learning*. Elsevier.

Radicchi, Filippo, Castellano, Claudio, Cecconi, Federico, Loreto, Vittorio, and Parisi, Domenico. 2004. Defining and identifying communities in networks. *Proceedings of the National Academy of Sciences*, **101**(9), 2658–2663.

Raup, David M. 1986. Biological extinction in earth history. *Science*, **231**(4745), 1528–1533.

Reed, William J. 2001. The Pareto, Zipf and other power laws. *Economics Letters*, **74**(1), 15–19.

Rice, John A. 2006. *Mathematical statistics and data analysis*. Cengage Learning.

Riley, Steven. 2007. Large-scale spatial-transmission models of infectious disease. *Science*, **316**(5829), 1298–1301.

Rish, Irina. 2001. An empirical study of the naive Bayes classifier. Pages 41–46 of: *IJCAI 2001 workshop on empirical methods in artificial intelligence*, vol. 3.

Roberts, David C, and Turcotte, Donald L. 1998. Fractality and self-organized criticality of wars. *Fractals*, **6**(04), 351–357.

Ročková, Veronika, and George, Edward I. 2018. The spike-and-slab lasso. *Journal of the American Statistical Association*, **113**(521), 431–444.

Rohe, Karl, Chatterjee, Sourav, and Yu, Bin. 2011. Spectral clustering and the high-dimensional stochastic blockmodel. *Annals of Statistics*, **39**(4), 1878–1915.

Rokach, Lior. 2010. Ensemble-based classifiers. *Artificial Intelligence Review*, **33**(1), 1–39.

Rokach, Lior, and Maimon, Oded. 2005. *Clustering methods*. Springer.

Ross, Sheldon M. 2014. *A first course in probability*. Pearson.

Rosser, J Barkley. 1999. On the complexities of complex economic dynamics. *Journal of Economic Perspectives*, **13**(4), 169–192.

Russell, Stuart, and Norvig, Peter. 2002. *Artificial intelligence: A modern approach*. Prentice Hall.

Sabidussi, Gert. 1966. The centrality index of a graph. *Psychometrika*, **31**(4), 581–603.

Sammut, Claude, and Webb, Geoffrey I. 2011. *Encyclopedia of machine learning*. Springer Science & Business Media.

Sapozhnikov, Victor B, and Foufoula-Georgiou, Efi. 1997. Experimental evidence of dynamic scaling and indications of self-organized criticality in braided rivers. *Water Resources Research*, **33**(8), 1983–1991.

Schapire, Robert E. 1990. The strength of weak learnability. *Machine Learning*, **5**(2), 197–227.

Schelling, Thomas C. 1971. Dynamic models of segregation. *Journal of Mathematical Sociology*, **1**(2), 143–186.

Schölkopf, Bernhard, Smola, Alexander J, Bach, Francis, et al. 2002. *Learning with kernels: Support vector machines, regularization, optimization, and beyond*. MIT Press.

Sen, Parongama, and Chakrabarti, Bikas K. 2013. *Sociophysics: An introduction*. Oxford University Press.

Sen, Parongama, Dasgupta, Subinay, Chatterjee, Arnab, et al. 2003. Small-world properties of the Indian railway network. *Physical Review E*, **67**(3), 036106.

Serrano, M. Ángeles, Boguná, Marián, and Vespignani, Alessandro. 2009. Extracting the multiscale backbone of complex weighted networks. *Proceedings of the National Academy of Sciences*, **106**(16), 6483–6488.

Sharma, Kiran, Gopalakrishnan, Balagopal, Chakrabarti, Anindya S, and Chakraborti, Anirban. 2017. Financial fluctuations anchored to economic fundamentals: A mesoscopic network approach. *Scientific Reports*, **7**(1), 8055, DOI: 10.1038/s41598-017-07758-9.

Shawe-Taylor, John, and Cristianini, Nello. 2004. *Kernel methods for pattern analysis*. Cambridge University Press.

Silva, Thiago Christiano, and Zhao, Liang. 2016. *Machine learning in complex networks*. Springer.

Simon, Herbert A. 1957. *Models of man: Social and rational*. Wiley.

Simon, Herbert A. 1991. The architecture of complexity. Pages 457–476 of: *Facets of systems science*. Springer.

Simon, Herbert Alexander. 1997. *Models of bounded rationality: Empirically grounded economic reason*. MIT Press.

Sims, Christopher A. 2003. Implications of rational inattention. *Journal of Monetary Economics*, **50**(3), 665–690.

Sinha, Ankur, Kedas, Satishwar, Kumar, Rishu, and Malo, Pekka. 2022. SEntFiN 1.0: Entity-aware sentiment analysis for financial news. *Journal of the Association for Information Science and Technology*, https://doi.org/10.1002/asi.24634.

Sinha, Sitabhra, Chatterjee, Arnab, Chakraborti, Anirban, and Chakrabarti, Bikas K. 2010. *Econophysics: An introduction*. John Wiley & Sons.

Song, Shunfeng, and Zhang, Kevin Honglin. 2002. Urbanisation and city size distribution in China. *Urban Studies*, **39**(12), 2317–2327.

Sornette, Didier. 2009. *Why stock markets crash*. Princeton University Press.

Sowell, Fallaw. 1992. Maximum likelihood estimation of stationary univariate fractionally integrated time series models. *Journal of Econometrics*, **53**(1–3), 165–188.

Spiegelhalter, David J, Best, Nicola G, Carlin, Bradley P, and Van der Linde, Angelika. 2014. The deviance information criterion: 12 years on. *Journal of the Royal Statistical Society: Series B*, **76**(3), 485–493.

Stanley, HE, Buldyrev, SV, Goldberger, AL, et al. 1993. Long-range power-law correlations in condensed matter physics and biophysics. *Physica A: Statistical Mechanics and its Applications*, **200**(1–4), 4–24.

Stauffer, Dietrich, and Sornette, Didier. 1999. Self-organized percolation model for stock market fluctuations. *Physica A*, **271**(3–4), 496–506.

Stock, James H, and Watson, Mark W. 2012. *Introduction to econometrics*. Pearson.

Stumpf, Michael PH, and Porter, Mason A. 2012. Critical truths about power laws. *Science*, **335**(6069), 665–666.

Sundberg, Rolf. 2019. *Statistical modelling by exponential families*. Cambridge University Press.

Sutton, Richard S, and Barto, Andrew G. 2018. *Reinforcement learning: An introduction*. MIT Press.

Suykens, Johan AK, and Vandewalle, Joos. 1999. Least squares support vector machine classifiers. *Neural Processing Letters*, **9**(3), 293–300.

Tan, Pang-Ning, Steinbach, Michael, and Kumar, Vipin. 2006. Classification: Basic concepts, decision trees, and model evaluation. *Introduction to Data Mining*, **1**, 145–205.

Thomson, David J. 1994. Jackknifing multiple-window spectra. Pages VI–73 of: *Proceedings of ICASSP'94. IEEE International Conference on Acoustics, Speech and Signal Processing*, vol. 6. IEEE.

Thurner, Stefan, Hanel, Rudolf, and Klimek, Peter. 2018. *Introduction to the theory of complex systems*. Oxford University Press.

Tibshirani, Robert. 1996. Regression shrinkage and selection via the lasso. *Journal of the Royal Statistical Society: Series B (Methodological)*, **58**(1), 267–288.

Tibshirani, Robert J, and Efron, Bradley. 1993. An introduction to the bootstrap. *Monographs on Statistics and applied probability*, **57**, 1–436.

Tong, Simon, and Chang, Edward. 2001. Support vector machine active learning for image retrieval. Pages 107–118 of: *Proceedings of the Ninth ACM international conference on Multimedia*.

Tong, Simon, and Koller, Daphne. 2001. Support vector machine active learning with applications to text classification. *Journal of Machine Learning Research*, **2**(Nov), 45–66.

Torgerson, Warren S. 1965. Multidimensional scaling of similarity. *Psychometrika*, **30**(4), 379–393.

Tran, Thanh N, Wehrens, Ron, and Buydens, Lutgarde MC. 2006. KNN-kernel density-based clustering for high-dimensional multivariate data. *Computational Statistics & Data Analysis*, **51**(2), 513–525.

Travers, Jeffrey, and Milgram, Stanley. 1967. The small world problem. *Psychology Today*, **1**(1), 61–67.

Travers, Jeffrey, and Milgram, Stanley. 1977. An experimental study of the small world problem. Pages 179–197 of: *Social Networks*. Elsevier.

Tsay, Ruey S. 2010. *Analysis of financial time series*. John Wiley & Sons.

Tsunoda, Tasaku. 2006. *Language endangerment and language revitalization: An introduction*. De Gruyter Mouton.

Tumminello, Michele, Aste, Tomaso, Di Matteo, Tiziana, and Mantegna, Rosario N. 2005. A tool for filtering information in complex systems. *Proceedings of the National Academy of Sciences*, **102**(30), 10421–10426.

Valdez, LD, Rêgo, HHA, Stanley, HE, Havlin, S, and Braunstein, LA. 2018. The role of bridge nodes between layers on epidemic spreading. *New Journal of Physics*, **20**(12), 125003.

Valente, Thomas W, Coronges, Kathryn, Lakon, Cynthia, and Costenbader, Elizabeth. 2008. How correlated are network centrality measures? *Connections (Toronto, Ont.)*, **28**(1), 16–26.

Vert, Jean-Philippe, Tsuda, Koji, and Schölkopf, Bernhard. 2004. A primer on kernel methods. *Kernel Methods in Computational Biology*, **47**, 35–70.

Wagstaff, Kiri, Cardie, Claire, Rogers, Seth, et al. 2001. Constrained k-means clustering with background knowledge. Pages 577–584 of: *ICML*, vol. 1.

Wang, Wen-Xu, Lai, Ying-Cheng, and Grebogi, Celso. 2016. Data based identification and prediction of nonlinear and complex dynamical systems. *Physics Reports*, **644**, 1–76.

Wasserman, Stanley, and Faust, Katherine. 1994. *Social network analysis: Methods and applications*. Cambridge University Press.

Watanabe, Sumio, and Opper, Manfred. 2010. Asymptotic equivalence of Bayes cross validation and widely applicable information criterion in singular learning theory. *Journal of Machine Learning Research*, **11**(12), 3571–3594.

Watts, Duncan J. 1999. *Small worlds: The dynamics of networks between order and randomness*. Princeton University Press.

Watts, Duncan J, and Strogatz, Steven H. 1998. Collective dynamics of "small-world" networks. *Nature*, **393**(6684), 440–442.

Wenzlhuemer, Roland. 2013. *Connecting the nineteenth-century world: The telegraph and globalization*. Cambridge University Press.

Weron, Rafał. 2002. Estimating long-range dependence: Finite sample properties and confidence intervals. *Physica A*, **312**(1–2), 285–299.

Wilson, Robert E, Gosling, Samuel D, and Graham, Lindsay T. 2012. A review of Facebook research in the social sciences. *Perspectives on Psychological Science*, **7**(3), 203–220.

Wolfram, Stephen. 1984. Cellular automata as models of complexity. *Nature*, **311**(5985), 419–424.

Wolfram, Stephen. 2002. *A new kind of science*. Wolfram Media.

Wolpert, David H. 1992. Stacked generalization. *Neural Networks*, **5**(2), 241–259.

Wood, Simon N. 2015. *Core statistics*. Cambridge University Press.

Wood, Simon N. 2017. *Generalized additive models: An introduction with R*. CRC Press.

Yamano, Takuya. 2001. Regulation effects on market with Bak–Sneppen model in high dimensions. *International Journal of Modern Physics C*, **12**(09), 1329–1333.

Yook, Soon-Hyung, Jeong, Hawoong, and Barabási, Albert-László. 2002. Modeling the Internet's large-scale topology. *Proceedings of the National Academy of Sciences*, **99**(21), 13382–13386.

Yuan, Jian, Ren, Yong, and Shan, Xiuming. 2000. Self-organized criticality in a computer network model. *Physical Review E*, **61**(2), 1067–1071.

Yule, G Udny. 1925. A mathematical theory of evolution based on the conclusions of Dr. J.C. Willis, F.R.S. *Journal of the Royal Statistical Society*, **88**(3), 433–436.

Yule, George Udny. 1900. VII. On the association of attributes in statistics: with illustrations from the material of the childhood society, &c. *Philosophical Transactions of the Royal Society of London. Series A, Containing Papers of a Mathematical or Physical Character*, **194**(252–261), 257–319.

Zhan, F Benjamin, and Noon, Charles E. 1998. Shortest path algorithms: An evaluation using real road networks. *Transportation Science*, **32**(1), 65–73.

Zhang, Menghan, and Gong, Tao. 2013. Principles of parametric estimation in modeling language competition. *Proceedings of the National Academy of Sciences*, **110**(24), 9698–9703.

Zhou, Aoying, Zhou, Shuigeng, Cao, Jing, Fan, Ye, and Hu, Yunfa. 2000. Approaches for scaling DBSCAN algorithm to large spatial databases. *Journal of Computer Science and Technology*, **15**(6), 509–526.

Zhou, Zhi-Hua. 2012. *Ensemble methods: Foundations and algorithms*. CRC press.
Ziqin, Wen. 2016. Zipf law analysis of urban scale in China. *Asian Journal of Social Science Studies*, 1(1), 53, DOI: https://doi.org/10.20849/ajsss.v1i1.21.
Zou, Hui. 2006. The adaptive lasso and its oracle properties. *Journal of the American statistical association*, **101**(476), 1418–1429.
Zou, Hui, and Hastie, Trevor. 2003. Regression shrinkage and selection via the elastic net, with applications to microarrays. *Journal of the Royal Statistical Society: Series B*, **67**, 301–320.

Index